New Infrastructures for Knowledge Production:
Understanding E-Science

Christine Hine
University of Surrey, UK

 Information Science Publishing

Hershey • London • Melbourne • Singapore

Acquisitions Editor:	Michelle Potter
Development Editor:	Kristin Roth
Senior Managing Editor:	Amanda Appicello
Managing Editor:	Jennifer Neidig
Copy Editor:	Mike Goldberg
Typesetter:	Diane Huskinson
Cover Design:	Lisa Tosheff
Printed at:	Integrated Book Technology

Published in the United States of America by
Information Science Publishing (an imprint of Idea Group Inc.)
701 E. Chocolate Avenue
Hershey PA 17033
Tel: 717-533-8845
Fax: 717-533-8661
E-mail: cust@idea-group.com
Web site: http://www.idea-group.com

and in the United Kingdom by
Information Science Publishing (an imprint of Idea Group Inc.)
3 Henrietta Street
Covent Garden
London WC2E 8LU
Tel: 44 20 7240 0856
Fax: 44 20 7379 0609
Web site: http://www.eurospanonline.com

Library of Congress Cataloging-in-Publication Data

New infrastructures for knowledge production : understanding E-science / Christine M. Hine, editor.
   p. cm.
  Summary: "This book is offers an overview of the practices and the technologies that are shaping the knowledge production of the future"--Provided by publisher.
  Includes bibliographical references and index.
  ISBN 1-59140-717-6 (hardcover) -- ISBN 1-59140-718-4 (softcover) -- ISBN 1-59140-719-2 (ebook)
  1. Internet--Technological innovations. 2. Computational grids (Computer systems). I. Hine, Christine.
  TK5105.875.I57N487 2006
  306.4'5--dc22
                              2005032107

British Cataloguing in Publication Data
A Cataloguing in Publication record for this book is available from the British Library.

All work contributed to this book is new, previously-unpublished material. The views expressed in this book are those of the authors, but not necessarily of the publisher.

New Infrastructures for Knowledge Production:
Understanding E-Science

Table of Contents

Section III: Prospects for Transformation

Foreword

In 2001, the UK Government launched a 5-year, £250M "e-science" research initiative. The term e-science was introduced by the then Director-General of Research Councils, Sir John Taylor, to encapsulate the technologies needed to support the collaborative, multidisciplinary research that was emerging in many fields of science. Such e-science research covers a wide range of different types and scales of collaboration. Particle physics, the community that developed the World Wide Web, now wishes to go beyond mere information sharing using static web sites for their new experimental collaborations at the Large Hadron Collider, now under construction at the CERN laboratory in Geneva. These collaborations typically involve over 100 institutions and over 2,000 physicists and engineers and are truly global in reach. Moreover, the physicists will be dealing with petabytes of data—a far larger amount of scientific data than scientists have hitherto had to manage, mine and manipulate. To handle such data challenges, the physicists will require a much more sophisticated set of shared computing and data services than currently offered by the World Wide Web.

The astronomy and earth sciences communities have a similar global reach but are more focussed on developing standards for interoperable data repositories than on computing cycles. In contrast to these global collaborations, biologists, chemists and engineers typically want to establish collaborations involving a small number of research groups and data repositories. Their requirements are for easy-to-use middleware that will allow them to set up secure and reliable "virtual organizations." Such middleware must assist researchers to routinely access resources and services at partner sites without having to memorize multiple passwords or manually negotiate complex firewalls.

In addition to the basic middleware to build such secure virtual organizations, these scientists require a powerful "Virtual Research Environment" that supports the needs of multidisciplinary research. Such an environment will consist of a set of sophisticated tools and technologies that will ease the extraction of information from data, and of knowledge from information. In the UK e-science program, for example, researchers in many projects are exploring the use of scientific workflows and knowledge management tools to support the scientists. Some projects are also evaluating the use of semantic web technologies in the context of these distributed collaborative organizations—the "Semantic Grid." It is also clear that these technologies will not only be useful to scientists and engineers but also to the social sciences and humanities. There are now an increasing number of projects exploring the way in which "e-science technologies" can be used to support social science and humanities research.

However, in addition to such work by practitioners of e-science, there is a complementary need to explore the sociological implications of these new collaborative technologies. I am therefore particularly pleased to see the publication of this collection of articles that begins an examination of these broader issues. I am convinced that such sociological issues will be as important as the technical ones in determining the uptake of e-science technologies and tools by the different research communities. I believe this collection will be an important contribution to our understanding of the potential of the new distributed knowledge infrastructure that is emerging.

Tony Hey
November 2005

Preface

Widespread international attention has recently been given to development of technologies to facilitate new ways of doing science. This book contributes to this burgeoning interest by using sociological methods and theories to explore the use of computers in scientific research. Specifically we analyze the increasingly prominent uses of information and communication technology (ICT) infrastructures for storing scientific data, performing analyzes and carrying out collaborative work, often known in the U.S. as cyberinfrastructures and in the UK as e-science. Led by data-intensive fields such as particle physics, astronomy, and genetics, new infrastructures are being designed that promise to allow scientific research to be conducted on a larger scale and with greater efficiency than previously conceivable, and to explore ever more complex questions. It remains to be seen how far this model will generalize, to what extent the models currently envisaged will translate to other disciplines and what the impacts may be for more traditional approaches. This collection looks at these innovations in the organization of scientific research, focusing on the factors which shape their inception, promote their uptake, lead to variability in application and result in the recognition of significant impacts. Our topic is, in short, the social dimensions of ICT-enabled science.

Social studies of science have established a set of approaches to the study of the scientific endeavour which explores processes of knowledge production as they emerge through particular social, spatial and material arrangements. Science and technology studies (STS) is an established field offering commentaries on scientific and technical developments, spanning the full range from top-level critique to detailed analysis and making contributions both to understanding the process of designing and to design itself. There has at times been an uneasy relationship between STS and the practitioners of science. The attitude of STS has sometimes been construed as anti-science, or as neglecting the concerns of scientists altogether in its focus on contesting philosophical accounts of scientific enquiry. This clearly does not exhaust the possibilities for the engagement of STS with science. A more hopeful scenario could be drawn from technology design, where approaches from STS have found popular acceptance. In technical spheres the relationship has often been less confrontational: It has become common for companies developing mass market technologies to employ ethnographers, using broadly STS-informed approaches, to inform their design activities. E-sciences, and more broadly the design of new infrastructures for the conduct of science, provide the occasion to extend this approach to the design of new technologies for science, allowing for STS to develop a much more constructive engagement with scientific practice. This is an unrivalled opportunity to demonstrate the relevance of social studies of science for the conduct of science itself.

An STS-informed understanding needs to be high on the agenda for those funding and designing new infrastructures for science, in order that technical capabilities be complemented with an in-depth understanding of social processes and consequences and with a theoretical tool kit to comprehend diversity. Innovation in the organization of scientific work should benefit from an enhanced understanding of social process. Woolgar and Coopmans in the first chapter of this book lay out a scope for the possible contributions that STS could make, but they note that the engagement of STS with e-science issues is in its infancy. That observation captures the motivation in putting this book together—to collect together and make more evident the ongoing work in the field, consolidating its contribution and making it more visible both to e-science practitioners and to the science studies community, beginning the work of forging relationships between STS analysis and e-science practice.

In the process of putting together this book it has proved helpful to adopt a fairly broad and flexible notion of appropriate technologies and situations to include. The goal has been, in part, to use the STS tradition of scepticism in order to question the way in which the e-science phenomenon is constituted. Part of our job has been to pursue connections which might otherwise have remained tacit between the current constitution of the phenomenon and the prefiguring technologies and policies, and to imagine the ways in which it might have been otherwise. A more practical reason for extending the scope of the technologies that we consider is in order to acknowledge the heritage of work on distributed, collaborative work and the foundations of scholarship on disciplinary practice and disciplinary difference. For e-science practitioners, it seemed that one of the most useful things we could do would be to bring this prior work into the domain of e-science and show how it can inform current dilemmas.

In the process there have inevitably been some omissions. Much of the existing work that we have been able to include has focused on communication and the preconditions for data sharing and other forms of distributed work. Much less focuses on computational aspects of the work of scientists, asking about changes in the knowledge production process and the ways in which findings are constituted, recognized and made communicable in e-science projects. Part of this is because e-science too is in its infancy, and thus examples to study for the whole of the knowledge production cycle are as yet rare. There are some promising signs that STS will be grappling with the import of computational e-science in emerging work on simulations and modeling (Lenhard et al., forthcoming 2006). Studying how knowledge is constituted has been the core of the sociology of scientific knowledge, and we can expect that this aspect of the e-science endeavour will be a focus in the future—all the more reason then to make sure that we are in on the ground now, exploring how the infrastructures that will support those knowledge production processes are being shaped.

I began this introduction by talking about e-science as an opportunity for STS to engage with science practice and policy on a constructive level. The opportunity is, however, not confined to an altruistic mode. There is a considerable pay-off for STS. Several of the chapters show that engagement with e-science is an opportunity to examine, refine and question accepted notions in STS. New technologies often provide reflexive opportunities, as much because they are perceived as new as for any specific transformative properties they offer. The reflexive opportunity that e-science offers to STS is the chance to explore knowledge production processes in the making, finding in their apparent novelty an occasion to reinvigorate established frameworks and highlight previously taken-for-granted assumptions about the ways that knowledge production works. The balance shifts between chapters, some focusing more on practical outcomes or policy critique for e-science, and some containing a stronger element of developing and reflecting on STS concepts.

The contribution to STS is methodological as well as theoretical. While social scientists have endeavoured to study scientific activity wherever it is carried out, the classic location for study is the laboratory. In the implementation of novel and spatially distributed ways of doing science we have an opportunity to see new locations of science in the making, and to ask new questions about the ways in which knowledge production is organized as a socio-technical process. We will learn new things about the processes by which scientific cultures change, about the importance of location, disciplinarity, and collaboration in science and about the development of new technologies in knowledge production contexts. E-science provides the opportunity to build on tried and tested methodologies of ethnography, of technology assessment and of scientometric and network analysis, and make them afresh for the situation in which they find themselves.

The first section of the book focuses on organized efforts to promote the development of new infrastructures, asking what motivates these efforts, what assumptions shape and constrain them and what structures they put in place. The impetus behind these examinations is to find modes of analysis that avoid falling into existing assumptions whilst maintaining a constructive dialogue. In the process, the chapters in this section highlight connections and limitations that might otherwise remain hidden between old and new technologies for science, and imagine alternative infrastructures that might serve the goals of different communities.

In the first chapter, Woolgar and Coopmans lay out a broad framework for the contribution that STS could make to the understanding of e-science. They argue that while STS approaches to e-science are in their infancy, there is a broad range of contributions to be made to understanding the genesis of these technologies, understanding the social and economic aspects of design, uptake and use and exploring the implications for the practice and outcomes of knowledge production. They argue that STS offers the potential for an in-depth examination of the ways in which notions of data, networks, and accountability develop in e-science contexts. For example, this makes it possible not just to follow the mobility of data but to ask how it is that data are recognized and rendered mobile and for whom. In a final section, Woolgar and Coopmans illustrate their argument by exploring notions of witnessing, with the aim of finding out how far e-science is replaying earlier debates about the adequacy of experimental reports. Notably, their argument suggests that we can use the juxtaposition of old and new versions of scientific witnessing to reflect on both forms, and, indeed, to question the distinction between old and new. E-science proves to be an opportunity to reflect on some previously taken-for-granted aspects of scientific communicative practice. Woolgar and Coopmans suggest that STS can both contribute to e-science and benefit from the engagement.

In the second chapter, Hine examines one discipline's use of information and communication technologies, exploring the dynamics which have produced a "computerization movement" within the discipline. The computerization movement framework highlights the cultural connections across diverse spheres that have been occasioned by computers, with the technology seen as a tool for beneficial social transformation. Hine argues that traces of such a movement can be discerned in the recent experience of systematics. This discipline has found itself with a high political profile thanks to its foundational role in biodiversity conservation efforts. This has produced a recent emphasis on using, and being seen to use, distributed databases to make data widely accessible. Grand visions for the transformation of the discipline have been promoted. The publicity accorded to grand visions allows systematists the opportunity to stress the progress that they are making, disaggregating and redefining the terms being used and taking care to point out that activities in this sphere need additional funding in order to succeed. Hine argues that the computerization movement in systematics, whilst not necessarily meeting with wholesale approval from those within the discipline, has nonetheless provided an opportunity for wide-ranging reflection and debate on its goals and practices. The upshot of this analysis for our understanding of e-science more broadly is that we should not be too daunted if grand visions fail to be realized. A payoff in terms of reflection, debate and learning across a spectrum of issues concerning technical infrastructure, practices and goals may occur quite independently of the realization of grander transformative schemes.

The third chapter turns to e-science initiatives to argue that, regardless of the broad gains to be expected from the stimulus they provide to develop technologies and explore visions, we should not adopt a laissez faire attitude towards their development nor assume that models will automatically diffuse across disciplines. Wouters and Beaulieu use the observation that science consists of diverse epistemic cultures (Knorr-Cetina, 1999) to explore how the locations in which e-science concepts are developed could be consequential for their subsequent generalizability. Indeed, Wouters and Beaulieu suggest that generalizable e-science might be an impossible dream, if we expect it to deliver infrastructures that travel across disciplines. Having established that much of the current focus of e-science initiatives is on computational work, they examine the epistemic culture of women's studies to suggest that quite different notions of data, analysis and infrastructure might prevail there. The idea of disciplinarity, and its consequences for the generalizability and deployment of new infrastructures, arises several times throughout this book. Wouters and Beaulieu provide the first of several discussions of disciplinary specificity, to be followed by Haythornthwaite et al., Merz and Fry in the second section and Nentwich and Barjak in the third.

The final chapter in this section moves to a different frame of analysis, to look at the ways in which e-science initiatives structure the labor relations between disciplines. Vann and Bowker introduce the term "epistemic IT," standing for the information technologies developed for use by scientists in knowledge production. Deliberately, this term includes both e-science and its predecessor attempts to promote the transformation of knowledge production through the development of new information technologies, since the chapter argues that preceding visions may have an unrecognized effect on the shaping of future possibilities. Vann and Bowker explore the predecessors of e-science to remind us that these visions are not new, and that they have the ability to shape goals, actions and expectations. In particular, they trace the emergence of a focus on interdisciplinary collaboration, and explore the ways in which this brought into being requirements for particular kinds of labor. Vann and Bowker show how grand visions of interdisciplinarity ask for the investment of labor, from domain scientists and computational experts, and describe one initiative where the commitment of scientists to make this investment was secured by some inventive funding arrangements. Vann and Bowker bring this section of the book to a close by exploring the "will to produce" that grand future visions promote, and the creative social and technical arrangements that may be required to enable that will to be pursued. In the next section the focus remains on the level of the experience of deploying new technical infrastructures, asking how they become embedded in and appropriate to their specific circumstances of use.

The second section of the book, on communication, disciplinarity and collaborative practice, asks what it takes for scientists to use new infrastructures for knowledge production. The focus shifts to the scientists using mediated communication for their collaborations and data exchanges, exploring the new forms of expertise that it demands and the extent to which certain disciplinary practices predispose their practitioners to use technologies in particular ways. These chapters suggest that use of new infrastructures can be a highly skilled form of work, and yet these skills may only be recognized when they break down, either in interdisciplinary initiatives or where the practical preconditions for their deployment are absent. This section focuses on the question of the specificity of new infrastructures for knowledge production as introduced in the first section by Wouters and Beaulieu, exploring in detail the dimensions of specificity and the constraints on extension of technologies beyond their contexts of production.

Merz begins the elaboration of the specificity of knowledge production infrastructures with a chapter that describes the embedding of electronic communications in disciplinary practice. She argues that the epistemic cultures of science can differ quite markedly in ways that are relevant for their adoption of electronic communications, and that the use of these technologies can develop in quite different ways in different communities. We should not, therefore, expect e-science to be a unified phenomenon. The argument is made by examination of a case study focusing on theoretical particle physics. Merz looks at the collaborative practices and preprinting conventions of this community, showing that use of electronic technologies is rooted in very specific ideas about the importance of collaboration and the nature of collaborative work. The relative freedom from physical location allows theoretical particle physics to be a particularly mobile community. Distributed collaboration making intensive use of e-mail is common, but does not render face-to-face communication redundant. Preprinting has a history which predates the availability of internet-based repositories, and is used by theoretical particle physicists in their everyday work in ways quite particular to their culture. Use of e-mail and of preprint archives is portrayed by Merz as highly cultural specific and deeply embedded in the culture of the discipline. This brings into question the extent to which models from this discipline might diffuse into other fields with quite different practices and expectations.

Merz shows that the communication technologies that a discipline uses are thoroughly embedded in practice. In the next chapter, Elvebakk explores another aspect of embedding, looking at the preconditions for data to be shared electronically. Through an exploration of the data-sharing practices of chemists, she shows how data have first to be rendered exchangeable in the eyes of their producers and users, by establishing the equivalence of representations with the material objects that precede them, and by making appropriate arrangements for the portrayal of contextual or otherwise tacit aspects of work with objects and data. This case study is situated within the existing literature on the relation of scientists with their objects of study, focusing on the processes through which objects are successively reconstituted through chains of representational practices. Elvebakk argues that current digital technologies are not radically different, instead continuing an existing trend diminishing the prominence of the material object.

If, as these two chapters have established, use of digital technologies takes place within distinctive disciplinary cultures, then it follows that interdisciplinary work can be particularly problematic. As Vann and Bowker discussed in the first section, interdisciplinary approaches have played an important part in the visions for new scientific infrastructures. In this section, Haythornthwaite, Lunsford, Bowker and Bruce discuss in depth the experience of distributed interdisciplinary work. They show that issues such teams face include the need to attend to work scheduling, learning practices, the nature of relations within and beyond the team and their use of technologies. While working practices in everyday discipline-based work often remain on a tacit level, working in a distributed interdisciplinary team can require that they be articulated and made explicit, so that all of those involved can share expectations and take care to head off likely pitfalls. New collaborative skills need to be developed to make distributed interdisciplinary work happen.

The final chapter in this section focuses again on disciplinary diversity, this time establishing a framework for making systematic comparisons between disciplines. Fry uses existing taxonomies of scientific disciplines to compare data on the experience of scientists from three disciplines with digital technologies. Interviews were conducted within high-energy physics, social/cultural geography, and corpus-based linguistics. Fry found that Whitley's (1984) framework accounting for organizational differences between scientific fields proved fruitful for understanding the different responses of these disciplines to the use of new technologies for collaboration and data sharing. Specifically, the framework highlights the importance of contrasting levels of interdependence between scientists and of varying levels of uncertainty around research problems, the objects of research, the techniques to be used and the evaluation of outcomes. Fry argues that we can use this framework to understand why some disciplines are more ready than others to adopt these new technologies, and to look at the way that the existing mechanisms of coordination and control of research practice may be affected by new infrastructures.

The third section of the book then moves to examination of the structures of science on a broader level, asking how new infrastructures might occasion wholesale change in the ways that science is communicated and in the inequalities between groups of scientists. The potential for change in the science system has, after all, been a feature of discussions about e-science and cyberinfrastructure, and it seems important to consider what signs are discernible at this early stage. Chapters in this section examine various aspects of the science system and of the structures and inequalities within it, to consider how far the promise of change is being realized.

In the first chapter of this section Nentwich examines the prospects for change in the scholarly publishing system. His approach is rooted in the tradition of Technology Assessment, a practically-oriented branch of STS that focuses on analyzing the potential consequences of emerging technologies, aiming to provide an informed basis for policy decision-making. Nentwich explores the evolution of new infrastructures for scholarly publishing and their differential diffusion across disciplines. He finds that there is likely to be considerable variation between disciplines, assessing potential change in terms of the endurance of print publishing in the face of electronic alternatives, the mechanisms of quality control, the economic and legal aspects of scholarly publishing and the varying collaborative structures of disciplines. He suggests that print systems may co-exist with new publishing infrastructures in the short term, but that print is ultimately unlikely to persist for most texts. There is a need, he argues, for a more organized constructive technology assessment to lead developments in desirable directions, involving representatives from the library community, from publishers, universities and, of course, researchers themselves.

In the second chapter in this section, Caldas assesses the emergent structures of science on the World Wide Web, analyzing the extent to which online structures mirror offline patterns of centrality and marginality. He draws on a previous study of the networks of communication and collaboration between European researchers of speech and language, and develops an analysis of the visible web presence of this community and the interlinkages between institutions. In scientometrics, STS has a strong heritage of exploring structures and evaluating differentiation within the science system using visible indicators such as publication and citation records. The World Wide Web provides the opportunity to develop webmetrics which deploy hyperlinks to evaluate emerging structures. Caldas finds that he can identify patterns within the web that map onto the offline structures of centrality and prestige. He also suggests that it may be possible to use the webmetric approach to identify "digital knowledge bases," emergent intensive knowledge zones that are particularly important and respected within a field. By focusing on the World Wide Web, Caldas is able to show us an emergent information infrastructure, not designed through any specific initiative but moulded out of the practices of individual institutions and researchers and their evaluations of one another. On this basis, we find that the new electronic infrastructures may mirror their more traditional counterparts.

Barjak, in the third chapter in this section, also evaluates emergent structures within the science system, and finds that there is a considerable continuity between old and new. His report is based upon a wide-ranging survey of internet use amongst European scientists, upon which multivariate analyses have been performed to explore differences between countries, disciplines, young scientists and more established researchers and between male and female. It has often been suggested that electronic technologies will have a leveling effect, enabling previously marginalized groups to participate more fully in society. In respect of science, it has been hoped that electronic technologies will enable fuller participation for less well-funded or less-experienced researchers and a broader accessibility of scientific information. Barjak finds, however, that levels of internet use are consistently reported as greater for male researchers and those at a more senior level. Rather than overcoming previous inequalities, Barjak suggests that the internet is tending to reproduce them, in a "hybrid divide" that encompasses both analogue and digital information.

In the final chapter we move to a focus on two aspects of the science system where inequalities have been starkly apparent, with an assessment of the implications of the internet for women scientists in developing countries. Palackal, Anderson, Miller and Shrum describe an interview-based study with scientists in Kerala, India, assessing the extent to which the internet is being used by women scientists and its impact on their experience and opportunities. Women are found to be generally highly restricted in their scientific careers in this patrifocal society, which both limits their ability to travel internationally and restricts their ability to network with male scientists locally. The internet, it is suggested, is allowing women to circumvent some of these restrictions, enabling them to connect with scientists internationally and to form research relationships which are less restricted by Kerala norms of gendered interaction. The internet is not as yet overcoming the restrictions which these scientists face, but is allowing them to circumvent some of their limitations and is a part of a growing awareness of the importance of international links among Kerala women scientists. This chapter demonstrates that the open networking possibilities provided by the internet can have a positive effect on the experience of previously marginalized researchers. When developing more sophisticated infrastructures in the future it may be important to try to preserve some of these advantages of their predecessors.

In summary, this book provides a highly differentiated account of the new infrastructures for knowledge production. Rather than having wholesale effects on the science system, the consequences are likely to be slow to emerge, and to be experienced very differently in different disciplines. The resulting structures are likely to have much in common with the existing science system, although some aspects of the new infrastructures may provide for new forms of participation amongst previously marginalized groups. How these changes evolve will depend very much on how debate and technological development are organized, and which groups are able to influence the agenda. STS perspectives suggest that it matters whose voice is heard when new technologies are being designed. They also suggest that users have a key role to play in deciding whether and how to adopt new technologies and in developing working practices which make sense of them. In the science community we have a highly empowered and reflexive group of people, who will ultimately be able to shape these technologies to suit their purposes. Hopefully, the analyses presented in this book will inform ongoing reflection and debate amongst policy makers, technology developers, scientists and STS researchers, enabling an enriched effort at designing the science system of the future.

Christine Hine

References

Knorr-Cetina, K. (1999). *Epistemic cultures: How the sciences make knowledge*. Cambridge, MA: Harvard University Press.

Lenhard, J., Küppers, G., & Shinn, T. (Eds.). (2006). Simulation: Pragmatic constructions of reality. *Sociology of the sciences yearbook*. Dordrecht: Kluwer Academic Publishers.

Whitley, R. (1984). *The intellectual and social organization of the sciences*. Oxford: Clarendon.

Acknowledgments

This book is a collective effort. I am very grateful to all of the authors for their willingness to contribute chapters, to learn from one another and to share constructive comments. I was conscious in putting the collection together that the contribution science and technology studies had to make to e-science was needed with some urgency: We argue often about the need to contribute to processes of technology design whilst they are in the making, and here we had an ideal opportunity to influence the practical and policy context of new infrastructures for knowledge production. For maximum effectiveness in reaching this audience, we needed to put the collection together fast. I am therefore especially grateful to the authors for their willingness to tackle challenging deadlines, and for their good grace and helpfulness in the face of pleas for rapid turnarounds. I also thank the staff at Idea Group Inc. for their efficiency in support of this process. My work on this volume has been supported by an ESRC research fellowship (Grant number R000271262-A). This book about new infrastructures for knowledge production would not have been possible without the generous assistance of the many scientists and members of the e-science community who have participated in interviews and helped STS researchers with ethnographic projects. Understanding the full ramifications of these new infrastructures as they develop depends upon this kind of engagement.

Christine Hine
Guildford, Surrey
November 2005

Section I

Framing New Infrastructures:
What, Why, How,
and for Whom?

Chapter I

Virtual Witnessing in a Virtual Age:
A Prospectus for Social Studies of E-Science

Steve Woolgar
University of Oxford, UK

Catelijne Coopmans
Imperial College London, UK

Abstract

Despite a substantial unfolding investment in Grid technologies (for the development of cyberinfrastructures or e-science), little is known about how, why and by whom these new technologies are being adopted or will be taken up. This chapter argues for the importance of addressing these questions from an STS (science and technology studies) perspective, which develops and maintains a working scepticism with respect to the claims and attributions of scientific and technical capacity. We identify three interconnected topics with particular salience for Grid technologies: data, networks, and accountability. The chapter provides an illustration of how these topics might be approached from an STS perspective, by revisiting the idea of "virtual witnessing"—a key idea in understanding the early emergence of criteria of adequacy in experiments and demonstrations at the birth of modern science—and by drawing upon preliminary interviews with prospective scientist users of Grid technologies. The chapter concludes that, against the temptation to represent the effects of new technologies on the growth of scientific knowledge as straightforward and determinate, e-scientists are immersed in structures of interlocking accountabilities which leave the effects uncertain.

Introduction

Despite a substantial unfolding investment in Grid technologies (for the development of cyberinfrastructures or e-science[1]), little is known about how, why and by whom these new technologies are being adopted or will be taken up. These questions are pressing, for at least three main reasons. Firstly, major current decisions about long-term investment in these technologies are effectively establishing modes of operation and use for many years to come. We need to understand now the ways in which we are fashioning the legacy for future outcomes and directions (Woolgar, 2003; cf. Wouters & Beaulieu, this volume). Secondly, it is widely agreed (even if sometimes only with the benefit of hindsight) that the inherent potential of new technologies does not itself guarantee their most appropriate uptake and use. Although much is said about the likely effects of Grid technologies, we need to understand what kinds of social circumstances facilitate and/or inhibit their use. Thirdly, we need to know about the uptake and use of the new technologies in order to discover the extent and the ways in which they are making a significant difference to the nature and practice of academic research, and to ensuing knowledge.

To elaborate this third point: It is almost a commonplace that significant changes in scientific direction and knowledge are associated with the development and use of new instruments and technologies. However, it is also well known that the nature and direction of change is unpredictable. For example, most would agree that many advances in knowledge of molecular structure are contingent on the development of electron microscopy (e.g., Barad, 1996). But, equally, it was in virtue of *not* possessing equipment for routinizing the analysis of observations of quasars that Cambridge radio astronomers were able to discover the wholly unexpected new astrophysical phenomenon of pulsars (Woolgar, 1978). These examples suggest we need to keep an open mind about whether and how Grid technologies might assist or inhibit the advance of scientific knowledge. The fact that it seems almost heretical, at this early stage of the game, to suggest that Grid technologies might do anything other than advance scientific knowledge, is itself an important feature of the emerging phenomenon.

This chapter aims to demonstrate the importance of addressing questions around the phenomena of Grid technologies and e-science from a science and technology studies (STS) perspective. In sections two and three, we indicate what an STS-informed approach to e-science might look like, identify key themes and topics for analysis and briefly show how these can be illuminated by bringing STS sensibilities to bear. The discussion here draws in part on the results of a wide-ranging consultative study carried out to identify and evaluate research themes and issues under the general rubric of "social shaping of e-science and of e-social science" (Woolgar, 2003)[2]. The fourth section then focuses on the nature of data under conditions of e-science. We revisit the idea of "virtual witnessing," a key concept in understanding the early emergence of criteria of adequacy in experiments and demonstration in 17th-century science. Our aim is to ask whether, and to what extent, the new Grid technologies facilitate significantly different forms of witnessing. Does virtual witnessing in a virtual age differ substantially from that undertaken at the origins of modern science? In our conclusion, we reiterate the main points of our argument and indicate two further directions for future research.

STS-Informed Approaches to E-Science

Two general issues in relation to Grid technologies provide a necessary starting point for outlining the questions that need to be addressed: the social and economic determinants of the design, uptake and use of these technologies; and the implications of the Grid for the nature and practice of natural and social science.

The Social and Economic Determinants of Design, Uptake and Use

An exploratory survery of existing "social shaping" perspectives on e-science and e-social science (Woolgar, 2003, section 4.8), showed that the social dynamics of the genesis of e-science had not yet been explored, and that there had been little attempt systematically to integrate social and economic questions around design, uptake, and use with ongoing initiatives in e-science. At the same time, increasing research effort is being directed toward general aspects of the genesis and impacts of information and communication technologies (ICTs), which has come to include an interest, expressed both in policy circles and by social researchers, in how ICTs might affect the scientific community (ETAN, 1999; OECD, 2000; Nentwich, 2003; for a critical perspective see Hine, 2002). This has generated a substantial body of results and research expertise, and key themes of previous work may have some applicability to the particular case of Grid technologies and academic research.

For example, recent research has shown that the uptake and use of new electronic technologies is generally unpredictable, counterintuitive and unevenly socially distributed (Woolgar, 2002b). Evidence from the research effort exploring the specific impacts of ICTs on scientific research suggests that uptake and use exhibit similar features, because scientific disciplines differ in terms of what they expect the new technologies to offer, and in terms of the social dynamics encountered when these technologies are put to use (Hine, 2002). This then raises the question of whether the same might be true in the particular case of Grid technologies as these are being introduced to benefit natural and social scientific research.

The Implications of Grid Technologies for the Nature and Practice of Natural and Social Science

Recent visions for e-science predict a radical transformation of scientific practices through a Grid infrastructure that enables high-speed and distributed, large-scale computations, and radically new ways of sharing large data repositories (Hey & Trefethen, 2002). The question of whether and how Grid technologies are making a significant difference to the nature and practice of academic research and to the knowledge which is generated require a critical analysis both of the nature and currency of these visions, and of the ways in which they are being enacted.

The project of addressing these three topics provides a happy confluence of two somewhat distinct social science research traditions. On the one hand, the social study of scientific knowledge has given us over 20 years of research into the effects of the social organization of science on the production of knowledge. On the other hand, more recent work under the auspices of "internet research," has shown the implications of new electronic technologies for a wide range of application areas. Curiously, these two traditions have enjoyed little engagement. Social studies of science have, with few exceptions, not taken as their focus the introduction of internet or Grid technologies. "Internet researchers" have by and large ignored the specifically scientific use and deployment of new internet-related technologies.

Science and technology studies (STS) comprise a large array of interrelated disciplines and overlapping intellectual themes. They involve contributions from sociology, anthropology, psychology, communications, and cultural studies; the main research themes draw upon intellectual currents as diverse as constructivism, reflexivity, actor network theory, feminist studies, and post-colonial theory. One subset of STS can be termed "social shaping" perspectives on science and technology (MacKenzie & Wajcman, 1985; Bijker & Law, 1992; Williams & Edge, 1996). "Social shaping" is defined very broadly to include all social scientific aspects of the genesis, use, implementation and effects of (new) technologies. While useful as a safeguard against technological determinism (Misa, 1988), it is sometimes associated with the assumption that "social factors" are somehow causally prior to emerging technologies (and that these social factors themselves exist independently of technological influences).

By contrast, recent, more sophisticated perspectives argue that the "social" and the "technical" are reciprocally elaborated: In other words, shaping occurs both ways. These latter perspectives suggest the need to consider bidirectional "impacts," for example, both how substantive research problems in the natural and social sciences shape the development and use of Grid technologies *and* how these technologies occasion the re-framing of research problems and/or methodologies. Relatedly, some existing "practice" studies of technology implementation make the point that in practice technologies are "localized," that is, they are made part of the local work routine of a group, while also changing that work routine in subtle or less subtle ways (e.g., Berg, 1997; Hughes et al., 2002). In other words, an understanding of Grid technologies in science should view them as both thoroughly situated in particular contexts of research practice, and yet highly consequential for the ways in which research is organized, conducted and communicated (Hine, 2003). This reinforces what we see as a central feature of a "social shaping" perspective on Grid technologies in science, namely, that it is necessary to interrogate the currency and meaning of what are perceived as the central components of e-science. The key is to develop and maintain a working scepticism with respect to the claims and attributions of scientific and technical capacity.

STS thus provides a wealth of existing analytic resources which may be drawn on for understanding the development and use of technology. However, it is not clear that the mere "application" of existing perspectives will suffice for an in-depth understanding of the specific ways in which the new Grid technologies affect the nature and practice of academic research. We need to remain open to the possibility that specific features of Grid technologies offer the opportunity for extending and developing our analytic frameworks. With these ends in mind, we identify three interconnected topics with particular salience for Grid technologies: data, networks, and accountability.

E-Science Topics for STS Analysis

Data

The collection, storage and exchange of enormous distributed and heterogeneous data resources has quickly become a central part of the vision for e-science. According to those running the UK E-Science Core Programme, "it is evident that e-science data generated from sensors, satellites, high-performance computer simulations, high-throughput devices, scientific images and so on will soon dwarf all of the scientific data collected in the whole history of scientific exploration" (Hey & Trefethen, 2003). Applications in the natural sciences such as astronomy, particle physics, bioinformatics, and weather prediction are thought to involve volumes of data expressed in "petabytes." Data are an important issue, not only because of the massive amounts expected, but also because complex constellations of numeric, textual and image data originating from a wide variety of sources make search engines, curation and access mechanisms complicated to design.

Social scientists studying the recently expanded process of electronic archiving for scientific research have suggested that this might be the start of a new relation between "data" and theory development, one that is different from the model exemplified by the traditional scientific paper (Bowker, 2000) and that crosses the boundary of the physical laboratory as the main site of knowledge production (Hine, in press). Bowker has indicated that data in scientific databases are increasingly expected to be of use to scientists from more than one discipline, to be reusable over time and to serve as the common basis for the development of scientific knowledge as well as political decision-making. The suggestion is that data collection, data storage, data mining and data sharing are no longer a means to an end (the "theory") but have become an end in themselves.

It is clear, then, that we need a detailed account of the practical processes through which the collection, storage, mining and sharing of data are envisaged and accomplished in the context of e-science. This includes, amongst other things, the emerging relationship between so-called "raw" data and data that are "prepared" for manipulation, sharing and presentation. The link between what is displayed (the data) and how it is displayed (presentation states) may be subject to increasing variation and fluctuation (Mackenzie, 2003). To what extent do new ways of visualizing and presenting data have implications for what counts as adequate "evidence?" Visions of data-sharing evoke a host of additional questions. How are a variety of qualitative and quantitative data in different formats and from numerous sources "socially negotiated" (MacKenzie, 1988) as shareable data that are useful for researchers from various disciplines? How do the emerging Grid technologies affect perceptions and practices of data-sharing in the sciences and social sciences?

The opportunity exists both to build on existing sociological analyses of the (changing) role of scientific data and critically to extend this work. The process of data collection, as well as the categorization of data, has been conceptualized as contingent and relative to the social relations within which this process unfolds, but once the data have entered an electronic database, they are then associated with particular, definite social effects (Bowker & Star, 1994; Bowker, 2000). However, as we mentioned above, a crucial finding of research into information and communication technologies in general is that their effects are often unexpected, unpredictable and unevenly socially distributed (Woolgar, 2002b), and there is preliminary evidence to suggest that the same may be true for digital databases (Hine, in press). Equally, the mobility of digital data is undertheorised, that is, it tends to be understood as an unproblematic, intrinsic attribute of digital data, a straightforward effect of the use of digital technologies for data generation and manipulation. However, particularly in an e-science context, it is important to examine in detail how expectations and perceptions of mobility are articulated and how the mobility of various kinds of data and resources is facilitated and managed in practice (for a preliminary attempt to do this, see Coopmans, in press).

Networks

Networks involve the linking and movement of data. A significant feature of the creation of working Grid technologies is the development of computer "networks" with the necessary bandwidth to cope with the demands of e-science, for example the linking of numerical data, images and text. At the same time, the term "network" is used to denote the creation of meaningful links between groups of researchers—both nationally and internationally, within and between academic disciplines—to facilitate effective communication and collaboration (Hey & Trefethen, 2002). Indeed, "lessons learned" from early Grid projects indicate that building and managing the latter type of network is proving to be at least as difficult and challenging as the former (and that, where it concerns features such as data protection and access control, "technical" and "social" understandings of networks are highly interdependent) (Jirotka et al., 2005).

This suggests we need critically to assess the various senses in which the concept of network has been used in the context of the development of science. Recent studies have investigated how new technologies enhance or weaken existing social networks (e.g., Haythornthwaite, 2002). A "networked perspective of innovation" is becoming increasingly popular in studies of business and organizational behavior (e.g., Powell et al., 1996; Hargadon, 2003). STS also has a long-standing interest in networks as a way of understanding how facts and artifacts come into being through the enrolment and juxtaposition of heterogeneous elements which need continual management (cf. Law & Hassard, 1999). However, it is important not to take the network for granted as an analytic concept. For example, Riles (2000) has shown how in particular circumstances, practitioners' evolving use of the concept of "network" rapidly becomes more sophisticated than the way in which it is used by academics. STS inclined perspectives on "networks" will need to include an interrogation of the use of the network concept over time and in different settings.

The concept of the network can also be used to address more traditional questions about communication, such as: Are patterns of communication between scientists likely to be significantly affected by the adoption and use of Grid technologies? Are users likely to be more or less sensitive to concerns about intellectual property rights (IPR)? Will the enhanced technical capacity for communication lead to different attitudes to risk, liability, and responsibility? Are changing reward and recognition practices affecting communication practices, and vice versa?

Finally, Grid technologies may lend themselves to innovative social science applications designed to trace various kinds of "connections." Some internet-based applications of this kind already exist, for example, the Issue Crawler project (Rogers, 2002). Yet we know from work on other electronic technologies that the existence of "offline" connections is crucial to the development and maintenance of "online" networks (cf. the third and fourth "rules of virtuality," Woolgar, 2002a; also Hine, 2000). Researchers travel between universities, to conferences and seminars and to meetings with developers and designers of the technology; they maintain contact with others via telephone calls (teleconferencing), and so on.

Accountability

We have said that a key question is whether or not Grid technologies will make a difference to the nature and practice of scientific research, and we have said that an interrogation of the currency and sense of the concepts of "data" and "networks" is central to this. At the same time, it is necessary to pursue this question in the wider context of changing conceptions of the utility and accountability of research. We suggest that the development and use of the new Grid technologies have to be understood in terms of the ways these draw upon and help to re-frame existing notions of "good," "valuable," or worthwhile research.

As is evident from published reports, press releases, committee meetings, project launches and similar events, the "wider context" has to include consideration of the involvement of major IT companies (such as IBM, Sun, Oracle, and CERN) and smaller software companies as partners (or prospective partners) in e-science projects. These materials make plain the hopes of the progenitors of e-science that the involvement of such partners will increase the chance that Grid technologies will become useful to others than academics (Hey & Trefethen, 2002). In what may be a relatively novel departure in scientific development, meetings of steering groups and management committees include consideration of appropriate "business models" for the future development of the technology.

In line with other reported major transformations in the accountability of knowledge (Gibbons et al., 1994), it is necessary to track possible changes in what counts as adequate knowledge and shifts in expectations about the utility of scientific research. This includes an investigation of arguments for the utility of Grid technologies in, for example, advancing national competitiveness, the national economy and public participation in research. A critical evaluation of these arguments needs to be based on monitoring their development over time and an examination of the social relations which promote their possibility. In other words, we need to examine the prospects for the utility of Grid technologies in the context of changing regimes of accountability, signified, for example, by an increasing demand for value-for-money.

In sum, this section has suggested that STS could be a fruitful basis for investigating how the new Grid technologies affect the nature and practice of science and vice versa. The key research questions include: How, in practice, are substantive natural and social scientific research problems made amenable to computer scientists' Grid solutions?[3] How do perceptions and expectations about the capacity of Grid technologies evolve over time? To what extent can we understand the nature and practice of academic research to be innovated by Grid technologies? Are new concepts of adequate or useful knowledge emerging in the process? How are new Grid technologies being "made at home" (Hughes et al., 2002) in particular contexts of research practice?[4] How, if at all, does this change what actually counts as "Grid technology" as new practices evolve? To what extent do new Grid technologies challenge existing social scientific knowledge about the social shaping of new technologies, including existing methodologies?

Characteristic of these questions, and of the particular flavor of the STS approach which we wish to advocate here, is the emphasis on the practical usage and enactment of Grid technologies. Of special importance is the need for research on practice which resists the temptation to adopt apparently self-evident features of the technical systems under study. The ethnographic sensibility of this style of STS is thus especially appropriate: Instead of merely taking them for granted, treat ideas such as "data" (including "data mobility") and "network" as the currency of discussions and actions of members of the e-science tribe.

Data and Its Mobility Under Conditions of E-Science: Virtual Witnessing in a Virtual Age

We have identified the general questions about e-science which can be analyzed from an STS-inclined perspective. In particular, we have argued that we need an STS framework which promotes a working scepticism with respect to what are perceived as the central components of e-science. What exactly is the currency and meaning of key terms such as data and networks and how can we understand them in relation to changing conceptions of the accountability of scientific research? In this section we illustrate this approach with specific reference to the enhanced manipulation of (especially visual) data which e-science promises.

An important analytic reference point for this discussion is the well-known argument by Shapin and Schaffer (1985; Shapin, 1984) about Robert Boyle's seminal contribution to the establishment of standards for communicating the results of scientific experiment. Boyle's 17th-century air-pump experiments are widely regarded as the foundations of the experimental method, which is still the model for scientific research today. A key component of the experimental method was the status of the scientific finding, or "matter of fact" as something that is distinct from causal explanations to account for it.

Matters of fact were the outcome of the process of having an empirical experience, warranting it to oneself, and assuring others that grounds for their belief were adequate. In that process a multiplication of the witnessing experience was fundamental. An experience, even of a rigidly controlled experimental performance, that one man alone witnessed was not adequate to make a matter of fact. If that experience could be extended to many, and in principle to all men, then the result could be constituted as a matter of fact. (Shapin & Schaffer, 1985, p. 25)

Shapin and Schaffer show that in order to connect the local performance of experiment to the production of universal knowledge (or scientific truth), the air-pump as a material technology had to be accompanied by a specific format of communication. Faced with the task of convincing observers who were not present at the scene of the experiment itself, Boyle advocated a set of communication protocols, a "literary technology" consisting of rules for description and witnessing, which effectively set the standard for the adequacy of scientific reporting. This technology included stipulations as to what was to count as sufficiently detailed illustrations, what kinds of prose and levels of description (including accounts of failure) would sufficiently convince absent readers, and so on. Shapin and Schaffer make the point that in establishing a communication protocol, Boyle also effectively establishes the standard for what counts as an adequate scientific experiment: "[W]hat counted as properly grounded knowledge generally, was an artefact of communication and whatever social forms were deemed necessary to sustain and enhance communication" (1985, p. 25). Scientific reporting is equivalent to making the claim that the experiment was indeed scientific. In virtue of Boyle's literary technology, observers removed from the scene of the actual experiment could nonetheless become its "virtual witnesses." The "technology of virtual witnessing involves the production in a reader's mind of such an image of an experimental scene that obviates the necessity for either direct witness or replication" (Shapin & Schaffer, 1985, p. 60).

In the terminology of recent claims about the effects of the internet and other electronic communication technologies, we see that Boyle was proposing nothing less than the death of distance (Cairncross, 1998). He was advocating a technology with radical effects, a technology which would establish a communications network whereby data could enjoy enhanced mobility and widespread sharing. At least two elements of Shapin and Schaffer's argument are highly relevant for investigations of e-science. First, the notion that communication, both its textual and social organization, can be understood as foundational to scientific knowledge. This includes a collective understanding of who has (and should have) access to experiments and findings. Second, the importance of the performance of boundaries between what is up for debate or dispute and what is not—in other words the allocation of a particular status to data, findings and theory which is specific to a certain community, place and time.

Shapin and Schaffer's analysis, however, differs in a very important respect from current attempts to investigate e-science. It rests on two sets of empirical observations: Boyle's attempts to establish the material, literary and social technologies foundational to the experimental method *and* the subsequent success of this method, key elements of which still constitute the core of modern science. Their way of accounting for that success differs from traditional stories about the self-evident superiority of the experimental method. By contrast, Shapin and Schaffer locate the success in the amalgam of the three technologies. Their argument is not quite as strong as to suggest that these three interrelated technologies form a *sufficient* (that is, determining) condition for the success of the experimental method. Nevertheless, in our reading of the argument, the coupling between Boyle's strategy for advancing the experimental method, and its subsequent success remains largely unexplored. At the time, of course, Boyle's success, versus that of Hobbes and other adversaries, was by no means guaranteed. It is only in retrospect that "the technology" becomes a candidate cause for the known outcome.

Shapin and Schaffer's account has been widely recognized as a key insight into the conditions which defined the nature of experimental science in the early modern period (e.g., Latour, 1993, p. 15ff.). But, most interesting for our purposes, many subsequent appropriations of the argument tend to adopt a summary version of the role of "the technology." This both implies a form of determinism and exacerbates the problem in the original which we just noted: What in detail accounts for the connections made between the deployment of "the technology" and subsequent "success"?

Kirby (2003, p. 235) notes that historians and sociologists have appropriated Shapin and Schaffer's concept and broadened its definition "to include any attempts to persuade others that they have witnessed a "natural" phenomenon without the need for them to actually witness the phenomenon directly." Kirby's own interest is in the ways in which fictional films "allow large sections of the public to "witness" phenomena without the need to directly witness these natural phenomena" (Kirby, 2003, p. 235). Other authors have enrolled Latour's (1990) notion of the "immutable mobile" to the general cause of this same argument. In a study of the functions of illustrations in six editions of Wilhelm Wundt's handbook for experimental psychology (1873-1911), Draaisma and De Rijcke (2001) argue that Wundt's extremely detailed schematics of new scientific instruments made a multiplication of the witnessing experience possible by virtue of their status as immutable mobiles: They gave such a "realistic" depiction of the instrument that it facilitated replication.[5] Based on this understanding of virtual witnessing, we would have to understand the Grid as furthering the cascade of inscriptions by making data more immutably mobile and thus potentially accessible to more, and more greatly dispersed, witnesses.

The main problem here is that elements of the original argument, and of its appropriation by subsequent writers, deploy a summary rendition of Boyle's activities and their effects. This tendency is encouraged, perhaps, by Shapin and Schaffer's use of the term "technology" as a collective term for Boyle's literary devices.[6] In this rendering, literary technologies, like immutable mobiles, come to have determinate properties and effects. Virtual witnessing is imagined to be a more or less straightforward outcome of literary technology. Immutable mobiles are imagined to provide the basis for long-distance exchange and transport of data. Yet is this not more than a little determinist? A central thrust of the analytic scepticism we have proposed as vital to STS treatments, is that these renditions be treated as *accounts* of technical capacity, and as *accounts* of the effects of literary technology and of immutable mobiles. Our argument is that the processes of genesis, attribution, negotiation and currency of (alleged) technical capacity should take centre stage. In line with this point, as we shall see, there are preliminary indications that the production, currency and use of the new Grid technologies does not at all guarantee the effects ascribed to them. For example, we need to ask to what extent and in what ways might the (literary, or Grid) technology be said to configure its users (Woolgar, 1991).

So what can we take from Boyle's promulgation of a new technology, itself widely heralded as a defining moment in the emergence of experimental science, for understanding the implications of current efforts to establish a Grid technology? We now address this question in two ways: Firstly, by comparing features of the account of Boyle's achievements with some recent work on the fate and implications of new electronic technologies; and secondly, by drawing upon materials from preliminary interviews and discussions with prospective users of the new Grid technologies, proto e-scientists in the fields of biological and biomedical research.

Data and Networks, the Virtual and the Real

Research into the effects of new ICTs has highlighted the ways in which anticipated outcomes and effects often contrast vividly with the uses to which they are put. In particular, this research suggests that claims about the new internet and other electronic communication technologies often give way to counterintuitive outcomes. As a general scheme for making sense of this phenomenon, Woolgar (2002a) organizes these research findings into five "rules of virtuality." These "rules" are intended as rules of thumb, or guidelines for making sense of claims about the effects of new communications technologies.

Woolgar's third "rule" is that "virtual technologies tend to supplement rather than substitute for existing (real) technologies." In other words, although new technology is often heralded as the displacement of existing practice, there is considerable evidence to suggest that new technologies sit alongside the continued use of "old" technologies. The iconic example, perhaps, is the much vaunted vision of the "paperless office." Far from expectation, the introduction of computer-mediated communication, e-mail, faxes and so on seems not to have led to less usage of traditional print and voice media. This prompts us to ask to what extent Boyle's advocacy of a literary technology rendered first hand witnessing obsolete (cf. Shapin & Schaffer, 1985, chapter 5). Did virtual witnessing come merely to sit alongside the continued use of the older form of witnessing? In what ways and to what extent was virtual witnessing regarded as a poor substitute for established practice? Similarly, how and why might the Grid come to be regarded as a poor substitute for "real" sharing of data?

Especially interesting for our purposes is the distinction in these discussions between, on the one hand, the activity or experience of actually witnessing an experiment (or data) and, on the other, the "virtual" (removed or secondhand) witnessing activity. Commentators are at pains to stress the difference between actual witnessing, at first hand, and merely "virtual" witnessing, with the latter imagined as a more or less adequate substitute for the former. Thus we find it said that the "aim of Boyle's lengthy experimental descriptions and engravings was to convince other scientific investigators of the validity of his experiments without the need for them to have actually 'witnessed' the experiments with their own eyes" (Kirby, 2003, p. 235).

Importantly, the emphasis on a contrast between "virtual" and "direct" witnessing leaves the character of direct witnessing unexamined. Or, more to the point, the very notion of "direct" witness itself thereby becomes reified. It becomes what its supposed virtual correlate is not. Moore (2003) makes an analogous point, albeit about a quite different area of technological application. She shows how discourse on the future of mobile telephony and of possible new future gizmos actually has the function of distributing and stabilizing acceptable descriptions of the current state of (mobile phone) affairs. By talking about the future, participants reaffirm conventional practice and understandings of the present. Similarly, we need to know, in relation both to Boyle and to our e-scientists, the extent to which discussions of the substitute for current practice (viz. the virtual technologies of witnessing and sharing), themselves solidify and congeal a specific understanding of what currently counts as witnessing and sharing.

We find in our discussions with e-scientists and in the recent literature a constant play between appearance and reality. For example:

The result is a visual simulation that looks and behaves much like the real heart it mimics. (Economist, 2001, p. 21)

The image you're looking at may appear to be a film of a living human heart. But it isn't. It's a highly sophisticated computer programme, using tens of millions of separate equations, combining graphics and mathematics to create one of the first of a new generation of simulations called "virtual organs. . . This programme simulates the complex activities of the heart and shows us just how a real heart would react to a range of conditions, physical and chemical. (Transcript from the Millenium Dome film commentary.)

Bringing this discussion back to our earlier focus on data and networks, we might ask how Grid and associated technologies are understood to extend or reconfigure the public space in which science is done and witnessed (Shapin & Schaffer, 1985, p. 39). The role of "raw" data here is especially interesting: In a move reminiscent of Boyle's decoupling of matters of fact and causal explanations, some scientific journals are already requesting that every scientific publication is accompanied by the release of the dataset its conclusions build on. As our e-scientists remarked:

If you can publish all the raw data in a public database under embargo that you make available to the reviewers of the paper when you submit it for publication to a journal, then they have access to far more on which to base their judgement of the veracity of the experimental work than if they only get the three pictures at the back of the paper. (DS14-12-2004)

Once all the data is available to everybody you can just go and look at somebody else's data and if you think what they've done with it is tosh and you can reinterpret it and get that published, then you can make [headway] yourself just because you've got access to that data and you never had that in the past and that democratises science. (DG04-01-2005)

If e-science can make it the case that a biologist anywhere in the world can access even the most complex of the models that people have developed, run it for themselves and be able to let that interact with the experimental work [] and indeed use it to help design what otherwise would be very compli-cated experiments you couldn't interpret, then it will have a big impact on what I see as a swing happening anyway [i.e., from a reductionist to a systems approach]. (DN16-12-2004)

This enthusiasm for making the data available sounds like a call for intensifying the virtual witnessing experience, or a return to the kind of witnessing-by-replication which Boyle advocated but which, as Shapin and Schaffer describe, in practice nobody succeeded in doing on the basis of his reporting alone. What our e-scientists are talking about appears to be the equivalent of Boyle handing over his pump. However, what exactly constitutes "raw data"? It turns out that these are not, in the words of one of our interviewees, the "dross" that is routinely produced by laboratories as part of their learning process. They are instead the carefully selected data that scientists choose to publish alongside their hypoth-eses. Similarly, the observation that the link between what is presented and how it is presented is increasingly unstable (Mackenzie, 2003) leads to a reimagining of the virtual versus the real: the real is now understood as the "raw data" to which access can be obtained, whereas the virtual becomes the representation of that data and its relations by various researchers for various audiences. One of our e-scientists illustrated this by saying that

[W]e [scientists] also can't interpret what we're doing just from the raw numbers that come churning out of the computer. Unless we take those gigabytes of information and convert them into images that a human being can understand and [use] the sequence in which these images change to produce the spectacular movies we produce, we can't understand what is happening. (DN16-12-2004)

The "raw" is thus represented as the new rock bottom point of reference. But, ironically, when e-scientists talk about raw data in this way, they often mean the data supplied by experimentalists. It was just these experimental matters of fact which Boyle contended needed witnessing!

There is another sense in which virtual witnessing "applies" and that is in relation to networks and the question of democratizing science. To be a virtual witness for seventeenth-century experimental science, one did not need an air-pump. The expensive apparatus only few could afford was—through Boyle's literary and social technologies—embedded in a system of assessment and verification designed to incorporate those that did not have the means to even be near an air-pump, let alone own one. Again, there are parallels with e-science. The vision crucially relies on the participation of those *not* involved in constructing the infrastructure for e-science. On the one hand, those who did not join the game from the beginning are unlikely now to get government funding for expensive computer facilities. On the other, there is an expectation that e-science will make it possible to "do top-class science" wherever you are, and that "small pockets of excellence around the world will be able to exist without having access to really expensive facilities—if they're really good and can convince others to let them join in the use of their expensive high-performance computer" (DG04-01-2005). But it remains to be seen how one becomes "really good" without having access to these facilities in the first place. Will e-science give rise to new networks, or reinforce the existing organization of science, including its geographical distribution?[7]

As we have noted, the advent of Grid technologies has been accompanied by considerable expectations about the greater sharing and mobility of data. It is perhaps worth noting that as part of these visions of the presumed benefits of Grid technologies, the notion of data sharing is posited as an unmitigatedly beneficial virtue for the operation of the scientific community and for science in general. We have attended some meetings where this apparently self-evident fact was reaffirmed by the invocation of descriptions of the Mertonian norms (notably, communalism) by which science supposedly operates. It is striking that this highly idealized, but now much discredited, version of science is implicated in advocacy of benefits of new technology for science. Quite apart from neglecting a large body of STS research on the details of scientific practice, this tendency has the unfortunate effect that attention becomes focused on the "barriers to take up" of the new technologies, as if the effects of the Grid would automatically swing into operation once we control for the pesky awkwardness of human psychology!

Accountability Relations in Virtual Witnessing

We earlier signalled the importance of accountability relations for understanding the apprehension and use of Grid technologies. Networks, data and mobility should be understood in terms of the accountability relations which they perform and enact, and from which they derive. It would be interesting to know much more about the kinds of accountability relations within which Boyle was operating. Shapin and Schaffer do comment that Boyle probably felt his own social standing needing downplaying, lest readers would think he was trading on his (relatively) noble station as a way of securing authority for his claims.

Relatedly, the issue of accountability relations in connection with the development of the new Grid technologies arises reflexively, as in the following example of deliberations about the place of social scientific research in the development of the Grid. The following is an excerpt from an e-mail message from the chair of a regional e-science centre, reporting upon the national director's deliberations about the potential contribution of social science.

At [the Director's] Technical Advisory Group (TAG) last week, he mentioned his interest in this area. Clearly, establishing a Grid infrastructure with applications which work, is his number one priority. However he made something like the following statement (phrased in a particular way on purpose) to TAG ... "he is interested in the human and social aspects of the Grid, and has had discussions with [two universities], and would like to know whether any members of TAG feel this is not appropriate." My interpretation is that he is keen, but wished to make sure that he would not be criticized for this interest in the future.

This message may be understood as a comment on the play of accountability relations within which the director is situated. A decision about funding social science investigations of the Grid is said to depend (in part) on the views of other members of the advisory group. The invitation to these members to express their views is framed by a statement of the "clear" priority to which they should attend. And, of course, the communication of the message itself performs and/or reinforces a certain set of expectations between its sender (the chair) and receivers.

Construed as a part of a structure of interlocking accountabilities, the complex currency and circulation of descriptions of the capacity and effects of e-science begin to make sense. One particular recent episode concerned the use by a journalist of (a preprint of) an academic social science report (Cummings & Kiesler, 2005) to support a point about "e-science." The journalist's article appeared in *The Chronicle of Higher Education* (Young, 2004) under the title "Does E-Science Work?," to which the journalist suggested the answer is "no." The academic report itself is based on a study of cross-disciplinary collaboration in science and ostensibly has little or nothing to do with "e-science."One of the authors commented:

As with any type of media coverage, one doesn't always have control over how the story is "framed"—our intention was not to focus on the issue of e-science but rather to focus on the issue of cross-institutional collaboration. The reporter was apparently more interested in e-science, because I don't remember the term actually coming up in our interview on the phone. (Cummings, 2004)[8]

The journalist seems to have used (the absence of) e-science as an opportunity for a newsworthy story—a story, in other words, which he judged might capture the interest of readers to whom he is accountable as a journalist.

The complexity of divergent relevant communities and audiences for e-science also begins to explain the often conflicting views expressed by our e-scientists. The e-scientists we interviewed revealed themselves to be ambivalent if not pretty sceptical about claims for the transformational qualities of the new Grid technologies.

[The Grid is] still a dream and a very important dream but I would not wish to emphasize as much as some of my colleagues have the extent to which the Grid is already helping us. I think there is quite a long way for the Grid people to go before they really transform the way in which we work. (DN16-12-2004)

It seems they find it necessary both to "support the hype" and yet to maintain a "realistic" view about the actual capabilities of the (as yet unproven) new technology. They spoke of the use of hype by prominent members of the community as an example where these participants "on the national scene" had to take into account their own political masters. The similar operation of a bifurcation of registers was noticeable during committee meetings, where we observed that, over time, participants progressively downplayed their previous enthusiasm for the prospects of the new Grid technologies.[9]

Participants thus appear to deploy and enact a multiple register for discussing and evaluating the new technology. This can be understood as a way of managing multiple accountabilities. Given their potential accountability to a wide range of audiences, colleagues, university authorities, current funders and potential future funders, we might conclude that proto e-scientists behave strategically. But such behavior may also be less deliberate than this characterization suggests. Part of the complexity is that it is not always clear, on any particular occasion, which audience(s) are relevant.

Does historical method tend to overlook those off-the-record remarks and activities available to ethnographic study, and which would otherwise give us insight into the existence and use of multiple registers? Construed, respectively, as inventor/propagator and users of a new communications technology, the activities of Boyle and his public seem extraordinarily decent and free of duplicity, especially by contrast with what we learn from more modern ethnographic accounts of the construction and use of new ICTs. The "grubbier" aspects of the advocacy and operation of literary technology receive rather little emphasis in Shapin and Schaffer's account. But were participants at the time themselves using a multiplicity of registers? If not, then we will have discovered an important difference between today's Grid-oriented scientific practice and the kind envisioned by Boyle. We would have to conclude that virtual witnessing in a virtual age requires scientists to be savvy strategists.

Conclusion

We began by emphasizing the necessity of understanding the unfolding institution and practice of e-science, given the likely enduring effects and implications of these investments for some time to come. This implies research which, for example, chronicles the emergence and impacts of Grid technologies using a longitudinal comparative design and/or which focuses in detail on the actual practices of the deployment and use of new Grid technologies. Whatever the focus of the research, we argued that it needs to embrace a form of analytic scepticism. Rather than adopting a "received view" of the central components of e-science, their currency and meaning need interrogation and analysis as part of an ethnographically-informed perspective. Among the key components associated with the likely success or failure of Grid technologies will be data, networks and accountability. But especially important to understanding these, is an appreciation that these are terms in flux. Their meaning and significance, just as those of the capacity and effects of the Grid technologies themselves, are to be understood as actively and interactively established.

Two further research opportunities are worth mentioning. First, we need to investigate apparent differences between e-science and e-social science, both in terms of the practical applications of Grid technology and in terms of their historical emergence—e-science being some years ahead of e-social science. We need to understand how the relationship between the two develops over time: To what extent can the new Grid technologies connect disciplines that have hitherto remained separate? Do the natural and the social sciences become more alike or do traditional differences become more pronounced with the development of practices such as data sharing and collaborative working over the Grid? Can we understand a possible convergence between the two as an indicator of the effectiveness of the new technologies?

Second, a common current assumption is that the use of Grid technologies for the conduct of social science ("applications") is necessarily distinct from the investigation of the social relations involved in the development and use of these technologies ("social shaping"). We need to investigate the possibility of developing an approach which integrates a "social shaping" perspective on e-science with a close involvement with the actual development and design of Grid applications for academic research. Collaborative work with developers of e-science applications would allow us to investigate the extent to which social-shaping perspectives can inform applications development and vice versa. Complementary to this, it is important to examine the potential utility of social-shaping approaches to applications of e-science. For example, the precise relation between ethnomethodologically-oriented workplace studies and science and technology studies needs to be investigated as e-science and e-social science applications evolve in relation to a variety of social-shaping approaches. As we have hinted, the alleged separation between the "technical" and the "people-centric" issues concerning the Grid (Hey & Trefethen, 2002), the boundary between these domains and the way people perceive and negotiate this boundary, may all rely on existing social and organizational structures of accountability which require interrogation. Our increasing knowledge of these aspects of the emergence and use of Grid technologies, while important in itself, should also provide an excellent basis for assessing the different ways in which social studies of science might be of value to e-science, and vice versa.

Acknowledgment

Our thanks for comments to Dan Neyland and Elena Simakova.

References

Barad, K. (1996). Meeting the universe halfway: Realism and social constructivism without contradiction. In L. H. Nelson & J. Nelson (Eds.), *Feminism, science and the philosophy of science* (pp. 161-194). Dordrecht: Kluwer.

Berg, M. (1997). *Rationalizing medical work: Decision-support techniques and medical practices.* Cambridge, MA: MIT Press.

Bijker, W., & Law, J. (Eds.). (1992). *Shaping technology/building society: Studies in socio-technical change.* Cambridge, MA: MIT Press.

Bowker, G., & Star, S. L. (1994). Knowledge and infrastructure in international information management: Problems of classification and coding. In L. Bud-Frierman (Ed.), *Information acumen: The understanding and use of knowledge in modern business* (pp. 187-213). London: Routledge.

Bowker, G. C. (2000). Biodiversity datadiversity. *Social Studies of Science, 30*(5), 643-83.

Cairncross, F. (1998). *The death of distance: How the communication revolution will change our lives.* London: Orion Business Books.

Coopmans, C. (2006). Making mammograms mobile: Suggestions for a sociology of data mobility. *Information, Communication & Society.*

Cummings, J. (2004). Personal communication.

Cummings, J., & Kiesler, S. (2005). Collaborative research across disciplinary and institutional boundaries. *Social Studies of Science, 35*(5), 703-722.

Draaisma, D., & De Rijcke, S. (2001). The graphic strategy: The uses and functions of illustrations in wundt's grundzuge. *History of the Human Sciences, 14*(1), 1-24.

ETAN Expert Working Group (1999). *Transforming European science through information and communication technologies: Challenges and opportunities of the digital age.* (Directorate General for Research, Directorate AP-RTD Actions: Policy Co-ordination and Strategy). Retrieved July 8, 2005, from http://www.cordis.lu/etan/src/document.htm

Fielding, N. (2003). Personal communication.

Gibbons, M., Limoges, C., Nowotny, H., Schwartzman, S., Scott, P., & Trow, M. (1994). *The new production of knowledge: The dynamics of science and research in contemporary societies.* London: Sage.

Hargadon, A. (2003). *How breakthroughs happen: The surprising truth about how companies innovate.* Boston: Harvard Business School Press.

Haythornthwaite, C. (2002). Strong, weak, and latent ties and the impact of new media. *Information Society, 18*(5), 385-401.

Hey, A. J. G., & Trefethen, A. E. (2003). The data deluge: An e-science perspective. In F. Berman, G. C. Fox, & A. J. G. Hey (Eds)., *Grid Computing—Making the Global Infrastructure a reality* (pp. 809-824). Chichester, UK: John Wiley & Sons.

Hey, T., & Trefethen, A. (2002). The UK e-science core programme and the Grid. *Future Generation Computer Systems, 18*(8), 1017-1031.

Hine, C. (2000). *Virtual ethnography.* Sage: London.

Hine, C. (2002). Cyberscience and social boundaries: The implications of laboratory talk on the internet. *Sociological Research Online, 7*(2). Retrieved July 8, 2005, from http://www.socresonline.org.uk/7/2/hine.html

Hine, C. (2003). *The ethnography of cyberscience: Information in contemporary science. (Research Proposal to the ESRC).* Retrieved July 8, 2005, from http://www.esrcsocietytoday.ac.uk/ESRCInfoCentre/

Hine, C. (in press). Databases as scientific instruments and their role in the ordering of scientific work. *Social Studies of Science.*

Hughes, J. A., Rouncefield, M., & Tolmie, P. (2002). The day-to-day work of standardisation: A sceptical note on the reliance on it in a retail bank. In S. Woolgar (Ed.), *Virtual society?—Technology, cyberbole, reality* (pp. 247-63). Oxford: Oxford University Press.

Jirotka, M., Procter, R., Hartswood, M., Slack, R., Simpson, A., Coopmans, C., Hinds, C., & Voss, A. (in press). Collaboration and trust in healthcare innovation: The eDiaMoND case study. *Computer Supported Cooperative Work (CSCW): The Journal of Collaborative Computing, 14*(4), 369-398.

Kirby, D. A. (2003). Science consultants, fictional films, and scientific practice. *Social Studies of Science, 33*(2), 231-268.

Latour, B. (1990). Drawing things together. In M. Lynch & S. Woolgar (Eds.), *Representation in scientific practice* (pp. 19-68). Cambridge: Cambridge University Press.

Latour, B. (1993). *We have never been modern.* London: Prentice Hall/ Harvester Wheatsheaf.

Law, J., & Hassard, J. (Eds.). (1999). *Actor network theory and after*. Oxford: Blackwell.

Mackenzie, A. (2003). These things called systems: Collective imaginings and infrastructural software. *Social Studies of Science, 33*(3), 365-387.

MacKenzie, D. (1988). The problem with "the facts": Nuclear weapons policy and the social negotiation of data. In R. Davidson & P. White (Eds.), *Information and government: Studies in the dynamics of policy-making* (pp. 232-51). Edinburgh: Edinburgh University Press.

MacKenzie, D., & Wajcman, J. (Eds.). (1985). *The social shaping of technology: How the refrigerator got its hum*. Milton Keynes: Open University Press.

Misa, T. J. (1988). How machines make history, and how historians (and others) help them to do so. *Science, Technology, & Human Values, 13*(3-4), 308-331.

Moore, K. (2003). *Versions of the future in relation to mobile communication technologies*. Unpublished doctoral dissertation, University of Surrey.

Nentwich, M. (2003). *Cyberscience: Research in the age of the internet*. Vienna: Austrian Academy of Sciences Press.

OECD (2000). *The global research village conference 2000: Access to publicly financed research.* (Netherlands Ministry of Education, Culture and Science, Directorate Research and Science Policy and the Organisation for Economic Co-operation and Development (OECD), Directorate for Science, Technology and Industry (DSTI), Committee for Scientific and Technological Policy (CSTP). Retrieved July 8, 2005, from http://www.oecd.org/dataoecd/21/8/1880819.pdf

Powell, W.A., Koput, K. W., & Smith-Doerr, L. (1996). Interorganizational collaboration and the locus of innovation: Networks of learning in biotechnology. *Administrative Science Quarterly, 41*(1), 116-145.

Riles, A. (2000). *The network inside out*. Ann Arbor: University of Michigan Press.

Rogers, R. (2002). The issuecrawler: The makings of live social science on the web. *EASST Review, 21*(3/4), 8-11. Retrieved July 8, 2005, from http://www.easst.net/review/sept2002/issuecrawler

Shapin, S. (1984). Pump and circumstance: Robert Boyle's literary technology. *Social Studies of Science, 14*(4), 481-520.

Shapin, S., & Schaffer, S. (1985). *Leviathan and the air pump: Hobbes, Boyle and the experimental life*. Princeton: Princeton University Press.

Technology Quarterly (2001). The heart of the matter. *Economist, 361*(8251), 21-23.

Williams, R., & Edge, D. (1996). The social shaping of technology. *Research Policy, 25*(6), 865-899.

Woolgar, S. (1978). *The emergence and growth of research areas in science with special reference to research on pulsars.* Unpublished doctoral dissertation. University of Cambridge.

Woolgar, S. (1991). Configuring the user—The case of usability trials. In J. Law (Ed.), *A sociology of monsters—Essays on power, technology and domination* (pp. 58-99). London: Routledge.

Woolgar, S. (2002a). Five rules of virtuality. In S. Woolgar (Ed.), *Virtual society?—Technology, cyberbole, reality* (pp. 1-22). Oxford: Oxford University Press.

Woolgar, S. (Ed.). (2002b). *Virtual Society? Technology, cyberbole, reality.* Oxford: Oxford University Press.

Woolgar, S. (2003). *Social shaping perspectives on e-science and e-social science: The case for research support.* (Economic and Social Research Council). Retrieved July 8, 2005, from http://www.sbs.ox.ac.uk/downloads/E-SocialScience.pdf

Young, J. E. (2004). Does e-science work? *The Chronicle of Higher Education, 51*(16), A25.

Young, J. E. (2005). Personal communication, January 11, 2005.

Endnotes

[1] Throughout this chapter, our discussion refers mainly to e-science as it is understood in the UK context.

[2] For the sake of brevity, we use e-science also to include e-social science throughout this chapter.

[3] It has been suggested that since the social sciences, in particular, are so disaggregated and non-cumulative, the kinds of change in the nature and practice of research and in the knowledge resulting from Grid technologies "might work on a very slow time-frame" (Fielding, 2003).

[4] And why does our text unthinkingly follow current conventions by deploying the capitalised form of the term "grid"?

5   Later in their article, the authors observe that the manufacturing of immutable mobiles was not a fast and frictionless process and that "Wundt's illustrations—like the three technologies they were part of—are best seen as facilitating factors in the distribution of a Leipzig-like psychology, certainly not as sufficient conditions" (Draaisma & De Rijcke, 2001, p. 20). Shapin and Schaffer, similarly, show the problem with replication.

6   Shapin and Schaffer use the word "technology" for machines, literary and social practices alike, in order to emphasise "that all three are knowledge-producing tools" (p. 25, note 4; also Shapin, 1984, p. 512, note 6).

7   On this point papers in the third section of this volume have some cautionary observations.

8   The journalist subsequently maintained: "I first heard of Mr Cumming's work when his collaborator, Sara Kiesler, gave a presentation that mentioned it at a conference. The paper they gave me that they have submitted to an academic journal is clearer in talking about e-science than is the report they published with NSF" (Young, 2005). There is only one mention of the term e-science in Cummings and Kiesler (2005).

9   Hine in this volume has some similar observations about strategic appropriations of future projections.

Chapter II

Computerization Movements and Scientific Disciplines:
The Reflexive Potential of New Technologies

Christine Hine
University of Surrey, UK

Abstract

This chapter examines some of the factors which help to create a momentum for developing new infrastructures for scientific research. Specifically it discusses the usefulness of the "computerization movement" perspective for understanding how innovations in scientific practice catch on and to what effect, arguing that we need to understand the role that wider cultural perceptions about the potential of new technologies play in shaping high level policy and day-to-day practice in science. A case study to develop this point is drawn from one scientific discipline, biological systematics. Examination of a recent policy document suggests that a computerization movement is in progress in this discipline, accompanied by a variety of strategic responses. It can be seen that a computerization movement in science can not only stimulate particular forms of technical activity, but also provide the occasion for focused discussions on the directions, goals and audiences for a discipline.

Introduction

It is axiomatic within science and technology studies that winning technological solutions are often not the ones which are "best" on purely technical grounds. Rather, technologies succeed through a complex set of social and economic considerations. One well-known example is the gas refrigerator, which lost out to electrical refrigeration. Gas-powered designs would have been quiet and reliable, but failed to win a strong market position thanks to the resources that General Electric was able to deploy to develop and promote their machines. Ruth Schwartz Cowan's (1985) account of the course of innovations in refrigeration is a classic in science and technology studies, demonstrating that we should not look to technical rationales alone to work out why a particular innovation succeeds.

In this chapter I will take this axiom of science and technology studies and explore its application to e-science. Specifically, I will use the momentum from Schwartz Cowan to justify looking at the dynamics which surround new technological infrastructures for research without assuming that technological superiority guarantees anything. I will be expecting, rather, that new technologies will need champions to introduce them to research communities and to demonstrate what they might be good for. I also expect that the technologies which are developed will be those which are credible within the community, desirable for individual researchers and institutions to invest themselves in and which are attractive to funding bodies. Whilst in the current volume we spend much of our time discussing the development of new, ideal and self-consciously innovative infrastructures for computing and data sharing, in this chapter I will be looking at factors which shape the landscape of existing projects within one discipline and the ways in which they are publicly promoted. This means that the technologies that I will be examining may not fit some of the more precise definitions of e-science and Grid computing. Indeed, one implication of my argument is that the precise definitions which emanate from high-level discussions are often subject to creative responses on the level of concrete initiatives, and at ground level it becomes harder to make sharp distinctions about the nature of the technologies involved.

I will be arguing that pronouncements about the role of information and communication technologies in science may be understood using frameworks developed to understand similar technologies in other organizational settings. Researchers of computerization within organizations have long argued that decisions to computerize are not taken on purely rational grounds, and that the impact of computerization is rarely predictable or incontrovertibly recognizable (Orlikowski, 2001). Rather, decisions to develop information and communication technologies may be taken on the basis of their fit with the ethos and image of the organization as represented by champions within the organization (Yates, 1989). These champions do not necessarily develop their ideas about technologies in isolation: They are likely to be local advocates of more widely-held sets of beliefs. This idea is captured in the concept of "computerization movements," as described by Iacono and Kling (2001), which consist of packages of values and expectations that develop around a technology, circulating in public discourse and appropriated within specific contexts of use.

Having first reviewed this literature for ideas about how computerization is enacted, and how its consequences emerge over time, I then evaluate their potential to illuminate a case study of one scientific discipline. The case study, of the discipline of systematics within biology, reveals that many of the observations from the literature on information technology within organizations are relevant here. Debates at policy level and within individual institutions revolve around both the practical qualities of new technologies and their symbolic qualities. The importance of being viewed as an up-to-date discipline and the topical importance of reaching out to users and providing accessible information have made the discipline particularly receptive to public discourses about the use of the internet. One key unanticipated benefit has been the processes of self-examination and of specification that the introduction of new technologies has occasioned.

Computerization as a Site of Anticipated Social Change

Computers have been a particularly potent site for imagining social change. This has been as true for organizational uses of information technology as it has for society at large. Yates (1989, 1994) provides historical accounts of the development of information processing in American firms in the 19th and 20th centuries. She argues that there were considerable variations in the uptake of these technologies between organizations, and that the differences could not be explained on purely functional grounds. She proposed a model to account for this variation which included three factors: the need for technologies based on the size and structure of firms; the availability of solutions for handling information;

and the prevailing ideology of the firm (Yates, 1994). She particularly stressed the role of managers as advocates for new technologies:

The technology alone was not enough—the vision to use it in new ways was needed as well. (Yates, 1989, p. 275)

A similar argument was made by Coopersmith (2001) when he suggested that fax communications did not diffuse without help, but needed advocates to encourage their use. This point is taken further by Orlikowski et al. (1995), who argue that even given the same technology, different organizations may make widely differing uses of it. This can happen when key advocates introduce a technology into an organization and mould the ways in which it is used.

Applied to scientific contexts, these points lead me to reflect that there may be an overall move towards development of new information and communication infrastructures for science, but that these technologies may be used very differently across scientific institutions, disciplines and research fields, depending on the ways in which they are introduced, the claims which are made for them and the models which are available to enable people to identify ways to use them. Examining the process by which new infrastructures for science are developed becomes key for understanding the subsequent uses of these technologies: both because of the biases which may be introduced into the technological designs (as Wouters and Beaulieu argue in this volume) and because of the learning which the communities go through in the development process. Also, perspectives from the history of organizational technologies stress the symbolic qualities of technologies. Often a technology will be introduced because of the values and expectations that it carries. In examining this issue, it becomes difficult to separate off scientific computing from a wider cultural sphere of values and expectations surrounding computing. Scientists may develop their own solutions, but they are likely to do so using discursive resources current in their broader cultural milieu. One key component of the argument that I make in this paper is that scientific computing is continuous with many of the beliefs about computers in the wider society, and in particular shares a belief that computers can be a source of social transformation.

The main framework that I employ in this paper to explore the momentum of beliefs about computing is the computerization movement (Kling & Iacono, 1988; Iacono & Kling, 1996, 2001). This notion builds on the idea of computers as potent symbols, and captures the dynamics which promote the computer as a key component of visions about social transformation. Rejecting the idea either that computers were adopted because they were clearly and rationally the best technology, or that everyone was being drawn in by marketing from powerful computer companies alone, the idea of computerization movements suggests that beliefs about the benefits of computers enjoy a wide cultural currency, manifested in more or less organized efforts to promote their use as a self-evident good. Kling and Iacono (1988) examined urban information systems, artificial intelligence, computer-based education, office automation, and personal computing, and found five ideological beliefs that they all shared: the centrality of computer technologies for reforming the world; improving computer technologies as a means to improving the world; that everyone gains and nobody loses from advances in computing; that more computing is better than less; and that the obstacles to these goals stem from people being difficult or obtuse.

The actual operation of the computerization movement works on three levels (Iacono & Kling, 2001). First of all, a core set of ideas about what the technology is and what the future vision is that it offers is arrived at. This technological action frame is then made available through public discourses, through government, the media and through organizations and professions. This is then appropriated at the local level: Often, here, the promise is not delivered, but these local difficulties are not necessarily recirculated back into public discourse. Kaplan (1995) deploys the computerization movement framework to examine the history of medical computing, and finds a prevalence of dreams and visions about the ability of computers to transform the field. She also finds that the visions rarely play out in practice by achieving all of the policy goals. However, rather than abandoning visions altogether in the face of these disappointments, the tendency is to revise the dream. The production of grand visions of transformation is too much of a standard practice in policy circles to be undermined by local instances of disappointment.

The notion of computerization movements has quite a different emphasis to the focus on advocates for new technologies described by Yates (1989): The first suggests developments across a broadly dispersed trend, whilst the latter implies the agency of charismatic individuals. These two aspects are, however, closely connected. In order to be persuasive it is likely that the ideas proposed by charismatic technology champions need to resonate with broader cultural currents. These currents position recipients to make sense of the ideas that champions are proposing. Computerization movements, on the other hand, need champions to make sense of them and interpret them for local consumption. A generally available idea that computerization is good needs to be related to a particular set of concerns held by a particular audience, and it needs to be demonstrated how a particular technology might help to achieve that goal. There is, then, a dynamic between dispersed movements and key individuals, and between broad concepts and more specific visions. It is this very complex set of dynamics which brings a computerization movement into being, renders it meaningful for audiences and makes it something that has a discernible (although often not predictable) impact on the experience of everyday life.

Computerization movements thus provide a way of accounting for the diffusion of similar sets of ideas about new technologies, whilst also allowing for local diversity and for the deviation of on-the-ground experience from grand vision. In the following section I will examine a recent event in which elements of a computerization movement are visible in recommendations about the role of computers in science. I will then go on to explore some ways of responding to these visions amongst those at the sharp end of practice in this field of science. I will argue that the computerization movement is a potent site for organizing activity and reflection, even though much of the local response resists the grand vision.

The "What on Earth?" Report

The case study which I explore comes from the recent history of systematics, the field of biology concerned with classifying and naming organisms and exploring the relationships between them. Systematics has made considerable use of information and communication technology in recent decades, increasingly depending on computerized methods for inferring groupings of organisms and hypothesizing evolutionary descent, and developing databases for cataloging specimen collections. Databases have also become significant as a means of communicating taxonomic information, and latterly there exist many online taxonomic information services including online identification guides, specimen collection catalogues and nomenclatural resources. Developments in computerization have been the site for debates about the scientific status of the discipline and its relations with users within biological science and beyond (Hine, 1995; Hagen, 2001). In recent years, the discipline has come under increased external scrutiny, thanks to its acknowledged status as the underpinning for attempts to conserve global biodiversity. The Convention on Biological Diversity instituted at the 1992 Rio Earth Summit used the term "taxonomic impediment" to capture the extent to which lack of taxonomic knowledge, and lack of access to taxonomic knowledge, was a global issue impeding conservation efforts. The taxonomic impediment has a specific geographic imbalance which has been a subject of political scrutiny, with expertise and specimen collections being concentrated in the more-developed countries and biodiversity hot spots often located in less-developed countries without the means to explore or protect their natural heritage.

It was against this backdrop that the key source of data I explore in this chapter was produced. In the UK in 2002, the House of Lords Select Committee on Science and Technology produced a report entitled, "What on Earth? The threat to the science underpinning conservation." In the following two sections I will first identify some components of the report that suggest a computerization movement in progress, and then move on to explore some reactions to this movement, in the evidence given by witnesses to the Select Committee, in the broader systematics literature around this time and in interviews that I conducted shortly afterwards with members of the UK systematics community. There is not space here to explore issues in depth: I will therefore concentrate on sketching core themes and providing selected illustrations.

Computerization Movements and the Visions for Systematics

The report was written to assess the current state of systematics in Britain, with particular reference to its capacity to fulfill the government's commitments under the Convention on Biological Diversity. Without explicitly setting out to examine the state of information and communication infrastructures, the Select Committee found themselves doing just this at every turn. In the one-page summary of the report, the role of digital information arises twice. First, under the topic of "increasing financial support," the report notes that an existing funding initiative could be usefully be focused in this direction:

We suggest that the Darwin Initiative should fund more projects to digitize UK collections in order to make more data available on the world-wide web and thus accessible to a larger number and variety of people. (Select Committee on Science and Technology, 2002a, p. 5)

The Darwin Initiative is a UK Government grant scheme focused on promotion of biodiversity conservation. While it is strongly linked to the Convention on Biological Diversity, it has no specific rubric to promote use of digital technologies in systematics. Instead, it is presented as "a small grants program that aims to promote biodiversity conservation and sustainable use of resources around the world" (Defra & ECTF, 2004). Many Darwin Initiative projects do involve provision of digital resources but the majority do not, at least as an explicit objective (only six of the 34 projects funded in Round 12 of the Darwin Initiative in 2004 explicitly mentioned production of a web site or database in their list of proposed activities). Nevertheless, the Select Committee were able to present digitization via the Darwin Initiative as an unproblematic means to achieve the UK's responsibilities to provide access to specimen collections. The link between digital resources and the accessibility of information is not confined to the Select Committee. The UK Department of Culture, Media, and Sport organized "The Designation Challenge Fund," offering money to museums for "imaginative and innovative schemes to boost access" (Resource: the Council for Museums Archives and Libraries, 2002). Of the 49 grants announced in July 2002, 30 made explicit mention of ICTs, involving online learning materials or access to digitized collections. Providing information online is now almost automatically counted as making it more accessible to a wide (but usually unspecified) audience.

Many in the systematics community have a much more nuanced understanding of the role of digital resources. It is recognized that images of specimens can serve a few of the functions of their material counterparts, but most systematic work is still thought to require access to the real thing. Also, an interactive digital identification key may be of little use when circumstances dictate that a more apt solution is a laminated sheet that can be taken into the field and used *in situ* without additional technological assistance. Access to digital resources may also be limited by cost and by reliability, and this can be a particular issue for the economically disadvantaged countries who are a central group of potential users. In particular, I was told that herbaria in such countries may have a strong preference for being supplied with high quality photographic prints of specimens rather than having access to digital images online, since the prints can be examined repeatedly and in detail offline and stored in cabinets in lieu of the actual specimens. The report promotes an idealized version of digital resources, which easily positions them as the solution to broader goals of making taxonomic information accessible, neglecting these more troubling aspects of their application.

On the topic of "collaborating and setting priorities," the report makes a more extensive comment on the role of information and communication technologies in the discipline:

We highlight the importance of digitizing the systematic biology collections, which will both increase accessibility of these data and help to update the archaic image of systematic biology. We also suggest that the systematic biology community should consider exploring new ways of presenting taxonomic information, in particular through increasing the amount of information available in digital form via the world-wide web, and should consider updating the system of naming previously undocumented species. (Select Committee on Science and Technology, 2002a, p. 5)

This summary statement captures the centrality which the report accords to the role of computerization in realizing a renewed systematics. In computerization movement style, computers are seen as central to achieving the new order, resulting in universal benefit, with more computerization incontrovertibly better.

The technological action frame of making systematics information available via the internet is promoted as leading to an accessible systematics which will automatically also gain more respect. Throughout the report it is clear that digitization is seen as a positive move, which will be received as such by diverse audiences of users, funders and potential systematists and which will improve the standing of the discipline.

Kling and Iacono (Iacono & Kling, 1988, 1996, 2001) point to the key role played by computerization movement organizations (CMOs) in publicizing the benefits of computerization, and specifically the way in which these organizations do framing work that shows people how they might use technologies in their own activities. The CMO provides a focus for the formulation and promotion of shared beliefs. While the Select Committee does a lot of framing work in their report, as we have seen, it only briefly takes on a role that would position it as a CMO. This report does advocate computerization, and this is also symptomatic of a broader pro-computerization ethos in contemporary government, but the remit of the Select Committee is of course not confined to computer-related topics. In preparing this report, the Select Committee gathered together and framed the work of other organizations, acting more as a temporary meta-CMO for other sites where ICTs are being advocated in systematics. Most prominent among the relevant organizations in contemporary systematics, and playing a major role in the report, is the Global Biodiversity Information Facility (GBIF).

GBIF could plausibly be represented as a CMO. It is a membership organization which operates through a governing board and a network of participating nodes. GBIF has an explicit campaigning role, its four key areas of emphasis being: data access and database interoperability; digitization of natural history collections; electronic catalogue of names of known organisms; and outreach and capacity building. As the statement of purpose in Figure 1 shows, the core of GBIF's ideology is the availability of biodiversity data, and the route to availability is believed to be the use of "digital technologies." The equation of availability with digitality admits of no exception. Just as Iacono and Kling (2001) found in other computerization movements, the benefits of computerization seem to be taken as self-evident and routine. While the Select Committee only briefly adopts the role of a CMO, GBIF functions in this way in a more sustained manner. GBIF, as an organization with substantial representation from within the systematics community, possibly also has a greater credibility as a source of vision within that community. However, as I will show in the following section, responses to the Select Committee's report and to GBIF amongst the systematics community have been somewhat equivocal.

The literature on the course of computerization in organizations made clear that technologies need to be made meaningful by advocates, in order to be recognized as appropriate solutions by organizations. Within computerization movements, champions advance the arguments for particular technologies and shape tech-

Figure 1. Description of GBIF's purpose (Retrieved March 8, 2005, from http://www.gbif.org/GBIF_org/bg1#whyneed)

Why is GBIF needed?

Good managers of natural resources and policy-makers know that their best decisions are based on results from the most accurate scientific analyses. Such analyses are based on solid, documentable data that have been recorded directly from the observation of nature. Such records are called 'primary' data.

Biodiversity is a handy, one-word name for all the species on the Earth, the genetic variety they possess, and the ecological systems in which they participate. Another way of thinking about biodiversity is as the "living resources" portion of "natural resources". A large part of the primary data on biodiversity are the 1.5-2.0 billion specimens held in natural history collections, as well as many geographical and ecological observations recorded by various means and stored in various media.

In making living resource policy and management choices, decision-makers are often forced to rely on analyses that are not based on primary data. This is because the world's store of primary data about biodiversity is not at present readily and easily accessible.

Future generations depend on the efforts made today to develop methods for sustainably using biodiversity. One very important part of the solution is rapidly, openly and freely delivering primary data about biodiversity to everyone in the global community, using digital technologies. Another part is ensuring that the primary data being collected today are stored in such a way that they will remain accessible to future generations.

nological action frames which show people what to do with them. In the Select Committee report, one champion's role is particularly apparent: Charles Godfray. At the time when the Select Committee's call for evidence was made, Godfray had published a paper calling for a radically new vision of systematics. Entitled: "How might more systematics be funded?" (Godfray, 2002a), the vision was at first published in the magazine of the Royal Entomological Society. It achieved a notoriety well beyond the entomological community, leading to a paper in *Nature* (Godfray, 2002b) and subsequent series of responses, and finally to a special issue of the Philosophical Transactions of the Royal Society edited by Godfray and one of his most vocal initial critics, Sandy Knapp of the Natural History Museum in London (Godfray & Knapp, 2004).

The Godfray "vision" was that systematics could reinvent itself and make itself more attractive for funding purposes if it moved towards a web-based model for communication. The vision was radical in that it required systematists officially to adopt the web as the medium for publishing a consensus classification: a fundamental shift in the operation of systematics away from the current mode in which competing classifications coexist and no ultimate arbiter decides which should be used. The web-based model would, it was argued, free systematists from existing chores of searching through centuries of published literature, and would make their work more readily available to users, providing a "product" which sponsors might be induced to support. Drawing on both functional arguments and the symbolic qualities of the technologies, Godfray explicitly envisioned the technology as a route to a transformed future for systematics.

Charles Godfray is not a systematist himself, and it might therefore seem surprising that he chose to promote a vision of what the discipline should do, and that he was treated as a credible commentator. He was, however, positioned to be heard in virtue of his expertise and his institutional position, which made him able to both make a credible and detailed case and to place that case in the wider arena. In terms of expertise, Godfray was able to position himself as an ally (or critical friend) of systematics by virtue of his amateur taxonomic interests and also to position himself as an intensive and influential user in his personal research and his role as Director of the NERC Centre for Population Biology. As a Fellow of the Royal Society, Godfray's word had considerable standing within the scientific community, and he was able to bring it to widespread attention. The timing of his statement was also fortuitous, in that it would have come to the attention of many of those responding to the call for evidence by the Select Committee at just the right time to make an impact.

The Godfray vision featured heavily in the report, and in the evidence given to the report. Systematics institutions giving evidence were asked about its desirability, and several user organizations proposed the web-based model in their responses to the call for evidence. While, as I will show in the next section, practicing systematists found the grand vision problematic, for many of their users and for the Select Committee it offered a neatly-packaged vision that appeared to resolve the problems of the discipline. Godfray therefore could be seen as an advocate for the computerization movement, developing a technological action frame that, at least for a certain audience, offered a plausible promise.

Strategic Responses to Grand Visions

Kling and Iacono (Iacono & Kling, 1988, 1996, 2001) suggest that while we might think that a movement will generate a countermovement almost automatically, this is not the case where computerization is concerned. Rather than opposing computerization per se, resistance tends to be focused on particular issues. Indeed, there are no obvious candidates for an organized counter-computerization movement in systematics. It is clear that in the current political climate, and in order to maintain funding, direct opposition to computerization would not be a viable strategy. The general positive air around computing means that resistance would probably be interpreted as yet another sign of the much-cited conservatism of systematics. There is, then, little prospect for resisting the trend, but there is ample scope for responses which aim either to disaggregate the phenomenon, or to harness the momentum of the movement in line with existing priorities of the systematics community. In this section I will examine some responses to the computerization movement in systematics, firstly in the evidence given to the Select Committee, then in the systematics literature of the time. While I do not quote directly from interviews, the analysis is informed by discussions with a number of systematists in the UK. These less formal interactions have been invaluable in exploring the nuances of the various responses to the "digital taxonomy" computerization movement.

The oral evidence given to the Select Committee shows considerable indications that the systematics community is making strategic responses to the computerization movement. I will quote extracts from one section of questioning at length, since it illustrates several facets of this broader dynamic. In this exchange, Lord Soulsby of Swaffham Prior is questioning Professor Chris Humphries, President of the Systematics Association. Humphries raises the topic of the Godfray vision with an approval tempered by an argument for the need to maintain existing media alongside new ones:

A recent position paper by Charles Godfray said that systematics should make use of the obvious benefits of e-science. This is all very well, and to a great extent he is making a play for the service end of systematics nomenclature identification, and those sorts of things. However, for many years to come there is going to be a definite need for monographs, floras, formal lists, checklists—all those kind of things—for the tropics, particularly, and for major monographs which can be used in laboratories and in the field at the same time. (Select Committee on Science and Technology, 2002b, p. 131)

Lord Soulsby responded with a question which pushed Professor Humphries to commit to whether he saw the need for the discipline to digitize for the sake of renewing itself, and its standing with other scientists. Professor Humphries responded with, again, a positive tone, but this time explaining that systematics institutions were already doing many of the things they were being asked to do:

Yes, I would accept that, but I would also defend the position that e-science is being rigorously pursued. There is a new organisation, called GBIF which is attempting to bring all the major databases of the globe together. I work at the Natural History Museum and in the Natural History Museum we have a whole section now devoted to writing the front ends of databases to go on to the internet. So, turning to genomics, the zoology department at the Natural History Museum has made a switch over the last ten years or so from vertebrate biology to invertebrate biology and, particularly, into process research, and this takes advantage of all the large databases containing sequence data as well as providing information to put into that database. So I think it is being embraced, and I think papers like Charles Godfray's are looking at one aspect. That was the point I was trying to make. (Select Committee on Science and Technology, 2002b, p. 131)

Lord Soulsby's next question pursued again the question of the image of the discipline and the need for vision. Professor Humphries again responded with an account of what was actually going on, suggestive of underlying problems not addressed by the grand vision. Current activities and future visions are aligned, but the suggestion is that the current activities, whilst exciting, are proceeding at measured and appropriate pace.

I think all monographs should venture to be on the web, as a matter of interest, but there are still problems with priority of names and that kind of thing. Again, in defence of what is actually going on, Norman MacLeod, the head of our Palaeontology Department at the Natural History Museum, runs a website which takes publications solely for publication on the web, involving palaeontological material. In terms of its look and what goes on, it looks very exciting to me. So I think what is happening is that there has been a response, but Charles Godfray, perhaps, is trying to push it faster than it is actually going. That is my impression. (Select Committee on Science and Technology, 2002b, p. 132)

This short interaction captures several features which recur in the response of the systematics community to the utopian computerization movement. Rather than straightforwardly rejecting the vision, systematists could be seen as making strategic responses which reject inappropriately universal solutions whilst aiming to show that they comply with the ethos of the vision. They are able to situate current activities within the broader concept of "e-science," by using a very flexible understanding of the term encompassing such varied activities as the provision of specimen databases on the internet, secondary publication of literature, experiments with virtual specimens and the use of computer programs for phylogenetic analysis. The questioning about e-science is therefore the opportunity for showcasing a range of existing activities. Responses employ the following strategies:

- stressing progress and positioning existing activities within the new agenda,
- disaggregating universal benefits to highlight sticking points, problems and the need for diverse solutions

- drawing on flexible definitions of technologies and of visions, and
- seizing the occasion to lobby for funding, stressing for example that the benefits of online taxonomy will only accrue if more funding is made available for the basic digitization effort.

The last of these points is not apparent in the evidence of Professor Humphries, but comes through very clearly in the evidence from the holders of systematics collections, fearful that visions of digital futures might sideline the importance of material collections, or neglect the need for extra funding to make digitization happen. Current funding opportunities were portrayed as overly focused on innovative technologies, without sufficient attention being given to the production of data to populate new infrastructures. Within an overall positive tone, responses were clearly measured, strategic and less than wholehearted. Ideas about the impression that particular technological solutions would make and the fundability of various options suffused the evidence given by systematics institutions and user organizations alike.

The reactions to the Select Committee's questions from the systematics community should be seen as a part of a broad political awareness that has seen systematics institutions in recent years face up to continued funding shortfalls by broadening their range of funding sources, making connections with new audiences and increasing efforts to promote a public image of relevance and responsibility. The systematics community, or at least the major systematics institutions, have had to become well practiced in this type of strategic response. Some of the exploitation of new technologies within the systematics institutions may be seen in this light, as initiatives which institutions have eagerly supported as the right kind of activity to be seen to engage in. This might seem to suggest that deployment of new technologies has been largely a cynical exercise, which is not my intention. Individuals within systematics institutions continue to engage in projects which make sense in terms of their scientific commitments. Private responses from systematists to the growing use of information and communications technologies vary between enthusiasm, resistance and resignation. Nonetheless, there has been a shift in the face of funding constraints in terms of the ways that the work of systematists is presented and promoted, and this has an inevitable effect on the kinds of work which are valued within institutions. Systematics institutions were therefore able to respond to a cultural landscape favoring e-science-style activities, by allying themselves with the overall trend whilst raising notes of caution.

A similar tone is discernible in the systematics literature around the same time. The publication of Charles Godfray's calls to systematists to seize the opportunities offered by the web, and their subsequent endorsement by the Select Committee, occasioned some consternation. A response from staff at the

Natural History Museum to the article in *Nature* shows the positive, but skeptical note in its title: "Taxonomy needs evolution, not revolution— Some changes are clearly necessary, but science cannot be replaced by informatics" (Knapp, Bateman, et al., 2002). Others took the opportunity to showcase existing developments (Bisby, Shimura, et al., 2002). Whilst largely accepting the overriding computerization movement ethos that more computing is necessarily better, the responses sought to show that culturally appropriate and acceptable solutions were indeed being developed, but at a pace that had little to do with grand visions of rapid transformation.

The question of culturally appropriate solutions arose again in a special issue of the Philosophical Transactions of the Royal Society, in which Godfray and Knapp came together to edit a range of contributions from users of taxonomy and practicing systematists on the future of the discipline, and in particular the role of new technologies. In their introduction, Godfray and Knapp naturalize the internet as the technology that taxonomists would have been using all along, had it only been available:

For example, taxonomists should continue to embrace the use of the internet and related media to enhance taxonomy, and fuse modern molecular approaches with traditional morphology. This is what Linnaeus and the great polymaths of the eighteenth century would have done if alive today. (Godfray & Knapp, 2004, p. 569)

This portrayal of systematics as already naturally innovative is in contrast to many of the prevailing characterizations of the discipline. It strikes a tone, however, with many of the contributions to the collection from systematists seeking to show that use of technology in the discipline is at the same time natural, and considered. Wheeler conducts an elaborate examination of what new cyberinfrastructures could and could not do for the discipline, but ends with a bold rallying cry:

Taxonomy, the sleeping giant, has awakened. Armed with the latest information technologies and a renewed sense of purpose, driven by the now or ever urgency of the biodiversity crisis, the community has the opportunity to lead a big science project that could define our generation. (Wheeler, 2004, p. 581)

Within the contributions to the special issue of the Philosophical Transactions of the Royal Society on "Taxonomy for the 21st century," it is possible to discern again the embrace of an overall computerization ethos at the same time as its aspirations are questioned and disaggregated. The initial Godfray vision becomes the occasion for a range of alternative visions to be articulated. A similar process of fragmentation can be seen in the fate of the Select Committee report, as it proceeded to official government response and then formal response from the Select Committee.

The Select Committee's report made a number of specific recommendations, amongst them a solution encompassing both GBIF and the Godfray vision:

We recommend the United Kingdom should take the lead and propose to the Global Biodiversity Information Facility (GBIF) that the GBIF run a pilot with some priority species to form the basis of a trial for Professor Godfray's suggestion of making taxonomy primarily digitized and web-based. A trial would demonstrate the benefits and pit-falls of this approach before implementing it more widely. (Select Committee on Science and Technology, 2002a, p. 6)

This recommendation was endorsed by the government in its response to the Select Committee report (at the same time as refusing calls to make substantial additional funding available). Following publication of the government's statement, reactions were invited from key institutions and organizations, and a formal response collated by the Select Committee. On the specific issue of the GBIF/Godfray vision in the report they noted:

12. The Government's response to this point was that they support the work of GBIF and has successfully promoted a pilot project in the light of Professor Godfray's recommendation. There was some division on this response from respondents. RBG Kew and the Royal Society were skeptical about GBIF taking on this role. The Royal Society believed that a web-based pilot should be undertaken by a major museum or botanic collection (or group of suchlike)... Several question the relationship of Professor Godfray's idea for a pilot project with GBIF funding—the Natural History Museum, when describing GBIF's calls for proposals said "It is not clear what, if any, connexion exists with Professor Godfray's ideas—our information suggests that GBIF have not adopted this approach at present." (Select Committee on Science and Technology, 2003)

This response is interesting for current purposes on two levels. Firstly, it illustrates the complex circulation of opinions and visions amongst organizations and institutions which comprises what might in Iacono and Kling's descriptions become a unitary computerization movement. Whilst the idea of the computerization movement is persuasive, there is still a need to remain attentive to the representation and re-representation of views and visions that this kind of public discourse entails. Secondly, it is clear that grand visions are fragile. What appeared in the Select Committee report to be a coherent computerization movement here fragments, as pieces of the vision are questioned and their ability to fit together is problematized, and the call for taxonomy to become web-based breaks up into more specific pilot initiatives and trials.

While there is little public commentary on GBIF and on the likelihood of achieving its goals of readily accessible biodiversity information, systematists were happy to offer opinions in interviews. Many of the people that I spoke with were involved in initiatives connected to the GBIF project. While they tended to see a lot of value in those individual initiatives, they were often wary or skeptical in relation to the overarching GBIF project. This is a community that has seen several grand initiatives come to little substance, and so while participation in this one is essential in case it just might be the one that succeeds, commitment to the grand vision is often resigned and guarded. Systematists want it to succeed, are prepared to make large personal commitments to making it succeed and yet have to protect themselves against the possibility that it may not succeed.

In evidence to the Select Committee's enquiry, in responses to Godfray's vision and in attitudes to GBIF it is possible to see the systematics community positioning itself in relation to an argument that is not amenable to outright rejection. Rather, they engage with calls for change by interpreting individual current activities and the discipline as a whole as being aligned with the proposed future direction, by disaggregating the phenomenon into a more diverse set of activities with varied audiences and by using the momentum and public prominence of debates to make renewed calls for respect and funding. The computerization movement thus becomes a resource for invigoration of debate, for examination of the issues of the discipline and for making activities visible within a wider arena. While a computerization movement might be seen as an imposition or nuisance, particularly when it originates outside the community as this one largely did, it can still be used strategically and become an asset. Systematics might not be renewed by the deployment of new technologies in quite the way that the computerization movement proposed, but it could certainly be said to have gained new vigor from the process of debate.

Conclusion: Computerization Movements as Resources

Systematics has recently been subject to a form of technological utopianism (Iacono & Kling, 1996) which sees web-based taxonomy as playing an essential role in an ideal future systematics. In the "What on Earth?" report, it is quite plausible to read the House of Lords Select Committee on Science and Technology as key advocates of a computerization movement. The Committee appear to share many of the beliefs about computing that Kling and Iacono identify in other movements: Specifically, that computerization is a universal good thing, and a tool of social transformation. While the Select Committee only temporarily, for the purposes of this report, acts as a computerization movement organization, other organizations which take on this role more consistently exist, most notably GBIF. Key advocates make the technologies meaningful for other participants, most notably in the case examined here Godfray with a vision of web-based taxonomy. A broad computerization movement within systematics currently exists, which sees digitization as a self-evident good thing for the image of the discipline and the availability of information to users.

Against the promotion of computerization as a self-evident good, any resistance amongst systematists could easily be seen as the result of misunderstanding, obstinacy or recalcitrance. There is thus no obvious counter-computerization movement as such within systematics. It is clear, however, that there is a large amount of scepticism amongst the systematics community as to whether many of the grander claims for computerization can be realized. Despite this scepticism, public responses to the computerization movement have been largely positive. The computerization movement provides an opportunity for the systematics community to mobilize extra political support, public approval and additional funding. Capitalizing on this opportunity involves some strategic responses to the formulation offered by the computerization movement. Specifically, strategies of response include: stressing progress and positioning existing activities against the new agenda; disaggregating broad statements about the benefits of digital technologies in terms of the functionalities of different media and the needs of specific user groups; and endorsing the vision and explaining the need for additional focused funding in order to achieve it.

Kling and Iacono therefore provide convincing ways to account for what is going on in the Select Committee Report. In particular, the concept of the computerization movement points to the presence of broad sets of expectations about computing which are realized, in specific instances, by advocates. Very similar sets of expectations prevail around computing, across different application domains and have done so for a number of years. The expectations that the Select Committee express about computing, and the Godfray vision in particular, are therefore to be viewed not as isolated instances, but as part of a broader cultural phenomenon. Viewing the influences on technological development in science in this way is a valuable reminder for the sociology of science. Science can usefully be viewed as continuous with its broader social setting in its beliefs about new technologies as much in the problems it seeks to address.

While the initial identification of a computerization movement is compelling, Kling and Iacono are less informative on what happens subsequent to the expression of these pro-computerization ideologies. Much of their discussion of the response to computerization movements revolves around the limited possibility of an organized countermovement. Within systematics there is no sign of an organized opposition that seeks to stand against computerization within the discipline: As Kling and Iacono suggest, such a countermovement would risk appearing unreasonably Luddite, such is the social currency of the pro-computing message. Instead of blanket counter-computerization movements, then, they expect reactions from groups attending to specific issues of potential concern.

It would be possible to stretch this account as a description of the reactions seen within systematics. Considerable debate was occasioned by the publication and subsequent high profile of the Godfray vision, and many commentators took the opportunity to point out why the vision was impractical or undesirable. The responses to the Select Committee report expressed concern that an inappropriate solution had been advocated. Thus far, the situation is as one might expect: not a blanket counter-computerization movement, but clear signs of resistance on specific issues, or from existing institutions who might be threatened by the proposed development. This would, however, do little justice to the complexity and strategic nature of many of the responses to the report. In the evidence given to the Select Committee, in the reactions to the report and in the literature surrounding the Godfray vision, we see a complex field of representations and re-representations, which involve a diversity of voices speaking for the technology and for one another. We also see that the momentum offered by the computerization movement is harnessed in diverse ways, and that without buying in to the total vision, members of the community can still find it a valuable resource. This is so both in respect of opportunities for increased public profile and funding, and in the provision of occasions to examine the direction and activities of the discipline on a strategic basis. Computerization movements, by the very nature of their grandness, evoke reactions that examine the status quo and explore feasible and desirable directions of change.

The current computerization movement influencing systematics is quite specifically focused on providing access to biodiversity information, including specimen databases and various taxonomic products via the internet. This particular vision of universally accessible information is consistent with the current political situation of systematics, stressing its importance in biodiversity conservation and the need to overcome geographical inequities in access to systematics expertise. While there is some discussion of e-science by systematists, the applications they discuss are quite different to the computationally intensive methods and large scale collaborations that most versions of e-science entail. It is for this reason that I distanced this paper in the introduction from e-science *sensu stricto*. On the basis of the case study presented here, I would suggest that disciplines will develop uses of information and communications technologies appropriate to the situations in which they find themselves, and that notions of e-science will be widely deployed for quite divergent technological practices. Appropriate technological solutions for a discipline may well also vary over time. While systematics is currently focused on provision of accessible information, in the past the focus has been on the scientific status of the work itself, and the attention has been placed much more on technological solutions to problems of taxonomic analysis (Hine, 1995; Hagen, 2001).

It might seem that by examining computerization movements one would be aiming to defuse their potency by showing how they work. This is far from my intention in this chapter, and it would be a futile exercise in any case. Computers will probably remain potent sites for imagining social transformation, at least until a new candidate comes along. However, even if we do not do away with the genre altogether, the analysis of strategic responses to grand visions that I present does suggest that computerization movements offer a considerable potential that could be made more of. If much of the potential of a computerization movement is in stimulating creative reactions against it, then more could usefully be made of occasions for expressing doubt and disaffection. This kind of debate should be encouraged rather than suppressed, if the potential of a movement to reinvigorate is to be realized. Some of this potential could be harnessed in the case of e-science, which can itself be understood as a computerization movement. There may be lasting doubt over whether radically new forms of scientific product and practice do indeed emerge from e-science initiatives. Still, through occasions when people disaggregate the notion of e-science and explore aspects that they do find desirable and feasible, there is a potential for new and interesting activity to be stimulated, whether or not it fits the strict e-science definitions. When evaluating outcomes it is important to look at the activity that has been inspired, as much as the specific product-based outcomes, or whether science has been transformed wholesale. Some less-anticipated side effects, such as the reflexivity which computerization initiatives promote and the collaborative working which they often stimulate, may have a significant benefit.

References

Bisby, F. A., Shimura, J., Ruggiero, M., Edwards, J., & Haeuser, C. (2002). Taxonomy, at the click of a mouse. *Nature, 418*(6896), 367-367.

Coopersmith, J. (2001). Texas politics and the fax revolution. In J. Yates & J. van Maanen (Eds.), *Information technology and organizational trans-*

formation: History, rhetoric and practice (pp. 59-85). Thousand Oaks, CA: Sage.

Defra and ECTF. (2004). *About the Darwin initiative.* Retrieved June 15, 2005, from http://www.darwin.gov.uk/about/

Godfray, C. (2002a). How might more systematics be funded? *Antenna, 26*(1), 11-17. Retrieved July 21, 2005 from http://www.cpb.bio.ic.ac.uk/staff/godfray/antenna_article.pdf

Godfray, H. C. J. (2002b). Challenges for taxonomy—The discipline will have to reinvent itself if it is to survive and flourish. *Nature, 417*(6884), 17-19.

Godfray, H. C. J., & Knapp, S. (2004). Taxonomy for the twenty-first century— Introduction. *Philosophical Transactions of the Royal Society of London Series B-Biological Sciences, 359*(1444), 559-569.

Hagen, J. B. (2001). The introduction of computers into systematic research in the United States during the 1960s. *Studies in the History and Philosophy of the Biological and Biomedical Sciences, 32,* 291-314.

Hine, C. (1995). Representations of information technology in disciplinary development—disappearing plants and invisible networks. *Science Technology & Human Values, 20*(1), 65-85.

Iacono, S., & Kling, R. (1996). Computerization movements and tales of technological utopianism. In R. Kling (Ed.), *Computerization and controversy: Value conflicts and social choices* (pp. 85-105). San Diego: Academic Press.

Iacono, S., & Kling, R. (2001). Computerization movements: The rise of the internet and distant forms of work. In J. Yates & J. van Maanen (Eds.), *Information technology and organizational transformation: History, rhetoric and practice* (pp. 93-135). Thousand Oaks, CA: Sage.

Kaplan, B. (1995). The computer prescription: Medical computing, public policy and views of history. *Science Technology & Human Values, 20*(1), 5-38.

Kling, R., & Iacono, S. (1988). The mobilization of support for computerization: The role of computerization movements. *Social Problems, 35*(3), 226-243.

Knapp, S., Bateman, S. R. M., et al. (2002). Taxonomy needs evolution, not revolution—Some changes are clearly necessary, but science cannot be replaced by informatics. *Nature, 419*(6907), 559-559.

Orlikowski, W. (2001). Improvising organizational transformation over time: A situated change perspective. In J. Yates & J. van Maanen (Eds.), *Information technology and organizational transformation: History, rhetoric and practice* (pp. 223-274). Thousand Oaks, CA: Sage.

Orlikowski, W., Yates, J., Okamura, K., & Fujimuto, M. (1995). Shaping electronic communication: The metastructuring of technology in the context of use. *Organization Science, 6*(4), 423-444.

Resource: The Council for Museums Archives and Libraries. (2002). *Tessa Blackstone announces £5.2 million grants for 49 museums around the country*. Retrieved February 23, 2004, from http://www.mla.gov.uk/news/press_article.asp?articleid=396

Schwartz Cowan, R. (1985). How the refrigerator got its hum. In D. Mackenzie & J. Wajcman, (Eds.), *The social shaping of technology* (pp. 201-218). Milton Keynes: Open University Press.

Select Committee on Science and Technology (2002a). *What on Earth? The threat to the science underpinning conservation*. London: House of Lords.

Select Committee on Science and Technology (2002b). *What on Earth? The threat to the science underpinning conservation. Evidence*. London: House of Lords.

Select Committee on Science and Technology (2003). *What on Earth? The threat to the science underpinning conservation. The Government's response and the Committee's commentary*. London: House of Lords.

Wheeler, Q. D. (2004). Taxonomic triage and the poverty of phylogeny. *Philosophical Transactions of the Royal Society of London Series B-Biological Science, 359*(1444), 571-583.

Yates, J. (1989). *Control through communication: The rise of system in American management*. Baltimore; London: Johns Hopkins University Press.

Yates, J. (1994). Evolving information use in firms, 1850-1920: Ideology and information techniques and technologies. In L. Bud-Frierman (Ed.), *Information acumen: The understanding and use of information in modern business* (pp. 26-50). London: Routledge.

Chapter III

Imagining E-Science beyond Computation

Paul Wouters
Royal Netherlands Academy of Arts and Sciences, The Netherlands

Anne Beaulieu
Royal Netherlands Academy of Arts and Sciences, The Netherlands

Abstract

This chapter problematizes the relation between the varied modes of knowledge production in the sciences and humanities, and the assumptions underlying the design of current e-science initiatives. Using the notion of "epistemic culture" to analyze various areas of scientific research practices, we show that current conceptions of e-science are firmly rooted in, and shaped by, computer science. This specificity limits the circulation of e-science approaches in other fields. We illustrate this using the case of women's studies, a contrasting epistemic culture. A view of e-science through the analytic lens of epistemic cultures therefore illustrates the limitations of e-science and its potential to be reinvented.

Introduction

The promise of technology and the dreams of what a tool might be good for are important in shaping its development and adoption. The futures of new technologies such as the Grid are not determined by the extrapolation of its technical possibilities. It is not even enough to postulate complex interactions between the social and the technical domain as the determinants of e-science. Rather, the future of e-science is, at least partly, created at this very moment, namely in the expectations with which the e-science enabling technologies are inscribed. The way proponents of e-science configure their dream can be analyzed as a "future script" (Brown & Michael, 2003). This script carries assumptions and presumptions that create boundaries between users and nonusers of the Grid and other e-science technologies. Moreover, the writing of the script itself already foreshadows these processes of social inclusion and exclusion by inviting some actors to coauthor the script and effectively excluding other voices. In this light, our chapter examines current concepts of e-science with the aim to uncover the foregrounding of certain future practices, and the backgrounding of others in this "practice of promise."

E-science is a particularly interesting case for the sociology of expectations because the creation of promise is a central feature of its current practice. The writing process of the future script includes the design of e-science web sites, the drafting of funding proposals and national programs as well as the creation of demonstrators and pilot projects. It is a very practical affair. And an open one—the nature of e-science as a dream about the future is not hidden but made quite explicit by its protagonists:

> . . . whereas the Web is a service for sharing information over the Internet, the Grid is a service for sharing computer power and data storage capacity over the Internet. The Grid goes well beyond simple communication between computers and aims ultimately to turn the global network of computers into one vast computational resource. That is _the dream_. But _the reality_ is that today, the Grid is a "work in progress", with the underlying technology still in a prototype phase, and being developed by hundreds of researchers and software engineers around the world. (CERN, 2005)

The dominant discourse about e-science is one of revolutionary changes in the way research will be conducted. In the words of the Grid Cafe: "The Grid is attracting a lot of interest because its future, even if still uncertain, is potentially revolutionary" (CERN, 2005). This dream hinges upon the difference between the web and the Grid. Whereas information-sharing is the core of the web, the sharing of computer power is the linchpin of the Grid. As is fitting for a dream, the exact features of e-science are not pinned down. What it means to practice e-science is not defined in any strict way. Rather, the proponents of e-science indicate what e-science will be by pointing to individual examples and case studies, usually in the form of pilot projects. This seems to be common to all e-science dreams, and generates a creative tension. We think that this tension is formative of the nature of e-science and should receive more attention from analysts and builders of this challenging new enterprise in the world of science and scholarship. We are interested in how this tension shapes the aspirations of e-science, both in the discourse about e-science and in research practices. Not least because it will determine whether important aspects of e-science infra-structures may become relevant to research and scholarship across the board or, alternatively, will be restricted to areas of computational research in the sciences and humanities.

We pose the following questions: To what extent is the dream of e-science constrained by the disciplinary background of e-scientists? What is the focus of current e-science projects? Is e-science truly a dream for everybody and a generic set of tools for all styles of scientific and scholarly research? And, last but not least, what would a non-computational e-science practice look like?

We discuss these questions by zooming in on the tension between generic description and specific embodiment of what e-science might be. We use the notion of epistemic cultures to highlight the different contexts of knowledge production. In the next section, we will define the analytical categories to investigate epistemic cultures in research in general and e-science in particular, drawing from the work of Knorr-Cetina (1999). We will then analyze the main e-science projects in the UK to find out whether e-science is indeed mostly about computation. The last two sections will make the case for the potential of non-computational e-science by using materials from fieldwork in women's studies with respect to e-science, and by bringing this together with a discussion of specific features of e-research in humanities.

Using Epistemic Cultures to Understand E-Science

Epistemic culture is an analytic framework, elaborated in the field of science and technology studies (STS) by Karin Knorr-Cetina (1999). This approach takes as a starting point that science is not a homogeneous set of practices but rather a patchwork of different ways of life. According to this framework, epistemic cultures can be characterized in terms of the objects being produced, the types of experiments conducted and the relations between units in a field. While developed to study experimental science, we are currently exploring the possibilities of adapting it to analyze knowledge production in other forms (Beaulieu et al., 2005; Wouters 2004; Wouters, et al., 2007 in preparation). This framework may therefore be a useful approach to study research across science, social science and humanities.

In this approach to science as culture, culture is defined as a shared set of recognizable patterned activities. Science is therefore analyzed not only in terms of material, organizational and praxiological aspects of knowledge production, but also in terms of symbolic and meaningful elements (Knorr-Cetina, 1999, p. 10). The notion of culture also highlights the development of researchers, and the way they maintain and adapt a particular culture that distinguishes them from other researchers. Each of the elements of an epistemic culture (its object, experiments and social relations) can help guide analysis of practices and of the way these change, for example, when there is an increased use of ICT. These elements are also useful in analyzing how ICTs are shaped by the cultures in which they are observed. Each element is briefly explained and illustrated below.

Objects

Every epistemic culture has a particular relation to the empirical. The ways in which an object of study is constituted therefore characterize an epistemic culture. For example, botanists constitute their object by collecting specimens in the field. In sociology, answers to surveys that give indications about behaviors may constitute the object of research. In each area of science, there are accepted ways of constituting objects and of manipulating them. These are situated practices, and not necessarily equivalent to textbook methodologies taught to undergraduates. They may also change. For example, in literary studies, the relation of researchers to their objects changes as texts are digitized. Whereas the traditional approach encounters a text in a relatively linear way, once digitized and tagged, the elements of a text can be mechanically recalled whenever the researcher wants. In other words, the relations to the empirical shifts as elements of the text are removed from the "field." In searching a data file, the elements of texts are removed from the book as the field where they occur. It is as though the encounter with the text is lifted out of the phenomenological encounter with the text in the act of reading, and replaced by an encounter with the text based on searching or other such manipulation. This kind of computer-based literary study constitutes the text as a new kind of object that can be studied in the "lab" of the analysis programs.

Experiments

A second analytic category of epistemic cultures is the way the object is interrogated, manipulated and made to yield interesting and unexpected results. To illustrate this with practices in literary studies, an example of an experimental strategy is the comparison of texts to determine authorship. Expertise for such comparisons might come from years of reading to familiarize oneself with a canon, with its boundaries and with the history of literary analysis of this canon. The analysis of texts then builds on the embodied knowledge of the reader, who interrogates the text based on this understanding. In a context where digitized texts form the empirical object, the interrogation of literary texts can take on a different form. For example, comparisons of vocabularies or structures may be based on the retrieval of similarities and differences based on automated search functions or on running the text through parsing software.

Relations between Units

Finally, the relations between units within a field are a key aspect of epistemic cultures. This can be seen as the "social organization" of a culture. The unit can be a research group but also an individual scholar. In literary studies, the "units" of this culture may in fact tend to be individuals, several of whom may be organized in relation to a common resource (collection, library, archive, national canon). In cases where special tools are being developed, scholars may be working in collaboration with other specialists and form research teams. This level of analysis addresses institutional issues in the production of knowledge (specialization, resource distribution) and other questions that have traditionally been addressed in the sociology of science (such as the economy of credit for work done, patterns of publication and recognition such as authorship, and the development of career paths).

The perspective of epistemic cultures, as an approach to understanding knowledge production, keeps sight of the fact that science is pursued not only by individuals or collections of individuals. Knowledge-making must be understood in terms of the material and symbolic dimensions needed to run experiments and communicate with others in the field. This notion helps to maintain an analytic stance that avoids technological determinism (the idea that technology determines social relations), that keeps sight of the contents and specificities of different types of work and that doesn't overly focus on the technical requirements of new tools.

A cultural approach to technology enables, moreover, a focus on meaning, on the locally elaborated practices around artifacts. It draws attention to the specific (as opposed to the taken-for-granted and unquestionable) ways in which technology can *become* significant, and maintains as a starting point that technologies are underdetermined. Given what we have said above about notions of promise of technologies as formative of practice, the epistemic cultures approach enables us to analyze the interaction of current and potential practices. We now turn to e-science initiatives and consider them in terms of the epistemic cultures from which they arise.

A Computational Dream

A common rhetorical strategy in documents and on web sites about e-science is to oppose the future Grid to the existing internet. In this way, a contrast is set up between a network of independent computers on the one hand (the internet) and a network working as one computer (the Grid). It is a computational dream:

It is practically impossible, nowadays, to do science without computers. Scientists are facing increasingly complicated problems which require much more than a blackboard! Often, a single computer, a cluster of standard computers or even a special-purpose supercomputer, is not enough for the calculations scientists really want to do. That's the way scientists are—always pushing the limits. Of course, computers are improving incredibly fast: processor power doubles every 18 months or so, a phenomenon often referred to as Moore's law. Still, they do not keep up with what scientists demand of them. As a result, scientists are often faced with situations where they "hit the wall", and which make it very difficult, very expensive, and sometimes downright impossible to achieve certain scientific goals with current computer technology. So some scientists started dreaming. They dreamt of a way to surmount these obstacles. They dreamt of having nearly infinite storage space so they would never have to worry where to put the data. They dreamt of having nearly infinite computing power available for their institution, whenever they need it. They dreamt of being able to collaborate with distant colleagues easily and efficiently, safely sharing with them resources, data, procedures and results. And, being always worried about their research grants, they dreamt of doing all this very cheaply—maybe even for free! (Dreaming costs nothing.) (CERN, 2005)

Note that the Grid is not only about computing, but also about data storage and distant collaboration. Nevertheless, computation is central as the key intellectual challenge driving the wish for e-science. Two ingredients are designated as the grounds for the need for more computational power: more complex problems on the one hand, and more data on the other. These two problems are intimately linked to the increased social scale of research: the need to coordinate resources for mega-projects across all continents. They emerge as central "motivators" for the Grid in most Whig histories (Butterfield, 1931) of the e-science communities[1]. However, the initial dynamics that generated the field of Grid computing, the core of present e-science initiatives, were quite different. To understand this, we need to delve into the history of the Grid.

It is a history of forward-looking documents and pilot projects in computer science that link to each other and build on each ancestor. The histories we have are mainly Whig histories: introductions or "historical" chapters in volumes aimed at promoting e-science and the Grid. Puzzles in computer science were the initial triggers of e-science. The first modern Grid emerged in 1995 with the I-WAY project. At a conference on supercomputing, researchers aggregated a national distributed test bed with over 17 sites networked together by a computer network, the vBNS. "Over 60 applications were developed for the conference and deployed over on the I-WAY, as well as a rudimentary Grid software structure to provide access, enforce security, coordinate resources and other activities" (Berman, Fox, & Hey, 2003 p. 13). Developing infrastructure and applications for the I-WAY seems to have been a transforming experience for the first generation of modern Grid researchers because they had to rethink their ideas about computer networks. "Whereas distributed computing generally focuses on addressing the problems of geographical separation, Grid research focuses on addressing the problems of integration and management of software" (Berman et al., 2003 p. 13). Thus, creating a global Grid was in fact the response to a local problem within a particular field in computer science, in other words, within the terms of a very specific epistemic culture.

Computer science is not the only driver, however. According to e-science proponents, a number of fields have been confronted with barriers in tackling complex computational problems and/or masses of data that threatened to swamp their computational resources. This is the case in astronomy, medical and cognitive science using digital imaging technologies and in many areas of the life sciences. But how far has this dream of e-science actually progressed? To what extent are researchers from specialties outside of computer science already involved in the construction of e-science practices? To answer this question, we analyzed the UK e-science program[2]. The UK program is one of the biggest e-science programs in the world. To be sure, the U.S. cyberinfrastructures programs organized by NSF predate the UK initiative. Nevertheless, the ambition to speak to scientific communities other than natural scientists has not developed until very recently. In the words of Stephen M. Griffin, a program director in the Division of Information, and Intelligent Systems at the National Science Foundation (NSF):

Historically, the USA Federal funding agencies have used narrow and restricted definitions of information/knowledge infrastructure—primarily focusing on computing and communications hardware systems, and more recently, networking middleware and scientific databases. Digital libraries research has reframed perspectives and dramatically broadened the IT applications spectrum. Innovative interdisciplinary applications in non-science, content-rich domains such as the arts, humanities and cultural heritage informatics are proving to pose altogether new and greater challenges for IT research and cyberinfrastructure development than have been encountered in computational science applications. (Griffin, 2005)

The UK program has taken the initiative to spread the gospel of e-science to the social sciences[3]. The first international conference on e-social science, for example, has been held in June 2005 in Manchester. Therefore, if e-science has evolved outside of its original context into other fields of academic research, this evolution should be visible in the UK e-science program[4].

The UK e-science program is run by a steering committee supported by a user group and a technical advisory group. The latter group is composed of 14 researchers with a background in computing and one bioinformatics expert. Computer science and physics are the dominant fields in the Steering Committee. The User Group is more diverse: its seven members are spread over six fields. Table 1 gives the composition of the Steering Committee and the User Group in

Table 1. Composition of committees UK e-science program

Steering Committee	
Field	**Number of members**
Computing	7
Physics	4
Bioinformatics	1
Neuroscience	1
Environmental science	1
Gerontology	1
DTI representative	1
Rolls Royce R&D	1
User Group	
Field	**Number of members**
Computational chemistry	2
Bioinformatics	1
Medical statistics	1
Physics	1
Meteorology	1
Astronomy	1

more detail. We must conclude that the e-science program is being run by experts from computational research fields only.

We also analyzed the projects running under the UK e-science program as of January 30, 2005. We looked at the project goal, the description of the actual research that the project entailed and the discipline to which the project contributed[5]. Interestingly, the three dimensions pointed to both the generic and field-specific features of e-science, the generic ones emphasizing the potential of e-science as a general enabling infrastructure and technology, the field-specific features underlining the local nature of technological and scientific puzzles. The project goal descriptions were most diversified and ranged from building the nuts and bolts of the Grid architecture to the design of racing yachts and new forms of publishing. This underlines that the e-science program is relevant to a wide variety of research projects[6]. If we look in more detail at the type of research actually foreseen, however, the picture changes. The development of data management and analytical tools is the largest cluster: One quarter of the research is devoted to data. Table 2 gives the types of research work that are most frequent[7].

The UK e-science research is mostly about data, data formats, data sharing and data analysis. In terms of the concept of epistemic culture: E-science seems especially relevant for the construction of the object. We also see that the infrastructure for research is strongly represented: Middleware, grid services and resource scheduling belong to the more frequently mentioned research activities. This refers to the relationships between units in the field. Visualization, distributed computing, and simulation are more directly linked to new research practices and may be interpreted as bearing upon forms of experimentation. Often they go together with the development of new data tools and these projects are almost exclusively devoted to quantitative research. This does not hold for the last type of research mentioned in the table: collaborative tools. These projects aim to build tools that in principle could be used in a variety of fields, although the fields in which they are tested in the first instance are mostly quantitative in nature. Forms of research that seem most relevant to non-computational research do appear, but they are scarce. For example, the creation

Table 2. Main research work in UK e-science projects

Type of research	Number of projects
Data tool development	62
Middleware	22
Visualization	20
Grid services	19
Distributed/shared computing	17
Simulation and modelling	16
Resource sharing scheduling/ brokering hosting	15
Collaborative tools	15

Table 3. Scientific fields figuring in UK e-science projects

Fields	Number of Projects
Computer science	93
Bioinformatics	36
Medical	21
Engineering	9
Environmental research	8
Bioscience	5
Physics	5
Chemistry	4
Astronomy	4
Social science	2
Neuroscience	1
Mathematics	1
Electromagnetics	1
Anthropology	1

of tools for textual analysis is mentioned three times, fieldwork support twice and browser development once.

This dominance of computational work is underlined by the relevance of the UK e-science projects to different scientific fields. Computer science is here clearly dominant, followed by bioinformatics. Table 3 gives the different fields to which the UK program aims to contribute in its projects.

As we can see, almost all fields are computational. Social science appears only twice, anthropology once and these projects may very well be mostly quantitative in nature. Of course, this in itself does not mean that e-science is being "colonized" by computational researchers. The dominance of computer science can easily be justified by the need to build sophisticated computer infrastructures that also pose new puzzles to computer science and engineering. With respect to infrastructure, the UK program is mainly focused on design. The relatively high number of projects that contribute to medical sciences is mainly due to projects that are creating new tools for clinical practice, such as fieldwork support and remote access to patient data. These same tools could arguably be relevant to, for example, the digital analysis of medieval manuscripts or the observational study of cultural behavior. In other words, there seems to be potential for e-science technologies as enabling technologies for a wide variety of fields.

Nevertheless, this potential is not yet instantiated as concrete possibilities. We can draw three conclusions from the analysis of the U.K. e-science program. First, most research in the e-science program aims to construct an infrastructure that is supposed to support all forms of e-science in the future. Second, the building of this infrastructure is related to and informed mainly by computationally-oriented research. The input from humanities has so far been virtually nonexistent and input from the social sciences scarce[8]. Third, this infrastructure-in-the-making has moved closer to actual research practices, which explains the dominance of data-oriented projects. Apparently, the basic computer infrastructure has developed far enough to be able to overlay this structure with data structures and middleware tools that should connect substantive research projects to the e-science infrastructure. This distinction between form and content is, we think, an interesting observation in itself. The conceptual move to create middleware that is at an intermediate level of specification with respect to the general information infrastructure on the one hand, and particular research practices on the other hand, points to a potential steering influence of e-science infrastructures. It is too early to tell whether and to what extent this standardizing potential will be actualized. Nevertheless, it seems worthwhile to follow its development. Furthermore, the ambitions of e-science projects underscore the importance of questioning the assumptions "black boxed" in digital infrastructure and tools.

Women's Studies

Epistemic culture as an analytic framework has mostly been applied to natural and life sciences. An important element that undergirds Knorr-Cetina's approach is the notion of experimental science, and its concomitant investment in the laboratory. While the configurations of object, experiment and relations may seem radically different in the humanities and social sciences, the framework of epistemic cultures can also be used to reflect on the practices in these fields— and to explore the limitations of the epistemic cultures framework.

For Knorr-Cetina, the laboratory is important in that it is the site where experiments and objects come together. The laboratory is therefore the place where an epistemic advantage can be gained, since the object does not have to be accommodated when and where it happens, in its environment. This element distinguishes, first of all, experimental science from "field" or observational science. As such, it would not seem to include practices in qualitative/phenomenological analysis, or many interpretative practices that are core activities of the humanities. Yet, western epistemology still has fairly directive requirements— disciplines must make their objects.[9] This always implies some removal from the field, even though a spatial configuration such as a laboratory may not always be the characteristic of such a removal.

Indeed, this broader view of what it means to draw on the empirical, to do experiments, provides insight into the way the promise of e-science might be received in certain epistemic cultures. It suggests which new scripts might be elaborated from these epistemic cultures in relation to e-science. In this discussion, we follow the road from promise to practice in opposite directions, and begin with practice in a specific epistemic culture. The material presented here is drawn from an ongoing project on a university-based women's studies group in a humanities faculty in the Netherlands. While the analysis is still in an early stage, some of our observations are used here to suggest how e-science might be dreamed otherwise.

Objects

In the women's studies group, the object of study is always highly contextual and reflexive. Scholars strive to understand their object in relation to its cultural provenance, and in relation to the theoretical and analytical tools used to examine this object. As stated in the dual research questions presented on their web site:

What representations of gender and ethnicity are currently being produced in theory, history, oral narrative, literature, television, cyberspace and film? What feminist methods, gender tools and analytic frameworks in relation to the power of dominant discourse are currently being produced as a form of resistance? (Women's Studies, 2005)

The object is therefore a cultural formation, in all its complexity, levels, variation and contexts (including that of interpretation). When encountering the web for example, it becomes an object of interest insofar as "cyberspace" provides the possibility of new forms of gendered representations, or new ways of interrogating them.

If the way women's studies constitutes its object were to be served by e-science tools, the following elements would be important to link practices and possible applications. Women's studies researchers would not tend to recognize their object as "data," which can be retrieved. Rather, the object to be accommodated by new technologically-supported practices would be the cultural representations of interest to women's studies. The possibility of recording, of displaying these cultural representations, would be more in line with the epistemic aspirations of women's studies. Furthermore, the possibility of showing and demonstrating digital and electronic settings would be valuable. An analogy with what e-science could do for women's studies might be the way the VCR was integrated into research practices. What is studied is not the videotape itself, but how the VCR does allow for capture, demonstration, close study and teaching of representations and cultural formations of interest to women's studies that may be present on television or in the cinema theatres. A similar way of "showing" the web or other ICT-mediated settings would likely be of interest.

Experiments

How can we speak of experiments on such objects? While experiments in the sense of "physical manipulation" are not particularly relevant, this notion can still be helpful to think about the ways in which objects of knowledge are constituted and interrogated. The representations analyzed are not solely a specific instantiation of a cultural form in a medium, but also the conditions that make this possible. The relations and conditions that support that instantiation are the object, and the discovery, analysis and expression of these constitute the process of knowledge-making in women's studies.

We are not talking, in the case of women's studies, about "digitization of empirical sources," at least not in the technical sense. The representations that are selected as objects of study may be in the realm of the digital, but they matter as "culturally" digital. This digitality is considered to be made up of many layers of practices, meanings and institutions, rather than solely as the expression of a bit of code on a particular type of hardware. To understand this object therefore requires an understanding of how it is created, valued and sustained.

Knorr-Cetina's classic example of the signal in high energy physics is perhaps not so different from the way women's studies make and understand their objects. In order to understand the signal, the accelerator must be understood in relation to all other events taking place in the experimental infrastructure. Similarly, women's studies tries to understand the context in which representations are made. In high-energy physics, the apparatus and context are highly technological (the infrastructure of the accelerator). The relation between the signal and the infrastructure is examined via complex mathematical models and calculations. In contrast, in women's studies it is the analyst herself who serves to "detect" the object of interest. The relations between the analyst and her object, as well as the context of this relation, are examined via sophisticated philosophical categories and reflexive processes. Working on an object in women's studies therefore means being aware of what is being made visible, of the role of the analytic framework for constituting and interrogating an object. The process of writing constitutes the main manipulation, the experimental setting where theoretical framework, object and interpretation come together through representation of the analysis in the shape of a text.

The analysis of "experiments," construed as the manipulation of objects by women's studies researchers, also highlights particular practices that have implications for what might be of value to researchers. Women's studies scholars highlight the importance of writing as a key knowledge-making activity, rather than as a way of "reporting" findings or disseminating results. Writing in women's studies involves maintaining an awareness of multi-layeredness of objects/interpretations/representations. Tools that support such activities might be imagined, beyond current office automation and word processing tools currently available. To elaborate a very simple example, when asked what would make her life easier, one researcher told of her need to keep track of what she has written. What she would like is an ICT application that enables her to keep track, not of her notes, nor of digitized archival material, but of the work she has done on a particular topic. This would enable her to answer the question: What have I said about such and such a historical figure? The idea here is to track meaningful passages, across a variety of texts. Currently, this can be solved by rereading one's work. This is not the same as managing data, because the context of the text within which this finds itself is important. This is one small way in which digitization must NOT mean removal from context, if it is to offer a meaningful function for scholars in this epistemic culture.

Social Relations between Units in a Field

As a philosophically and literary studies-oriented group, it might be expected that very individualistic modes of operation would be privileged. Indeed, if one looks at elements such as publications, there are rarely coauthored works.

In order to maintain an institutional space where this work can proceed, however, strong social relations are needed. In the case of this particular group, there is an important interaction between the use of e-mail and web-based communications, the notion of "internationalization," and the securing of a space for the pursuit of women's studies in the university. Women's studies practitioners were already attuned to the international scene, often explaining that women's studies could not be viable in terms of the national context of their original discipline. Very briefly, the notion of internationalization is significant because in the past decade, it has become an important criterion of scientific quality in the specific national context. E-mail communications therefore are a way of maintaining and expanding this international connection, while web sites can serve to display this international orientation to local administrators in the university, for example.

In this particular fieldwork to date, the activity most mentioned (and usually mentioned first) in relation to ICTs is sustaining of networks. Contacts with faraway colleagues and groups sustained by individual e-mail, mailing lists and web sites are considered crucial to the success and activities of the group. This plays out in a number of registers. For example, contacts serve to develop work and ideas for members of a community that is very small in the national context. Awareness of conferences, workshop and publishing opportunities also happens via e-mail and mailing lists. Making contacts that are essential for the organization of group or bilateral activities and for obtaining grants also happens via e-mail. This network enables obtaining funding, or holding events (PhD summer schools) that would not be available or feasible on a national scale. Related to this networking activity, the "international" dimension functions as a mode of legitimation in the national context where this fieldwork is taking place. Having international links is considered a criterion of excellence, one which researchers in this group have been able to fulfil very effectively. As a consequence, the display of international contacts as forms of legitimation, in the face of university administrators, other departments in the university or other scholars from the mother discipline is also performed via web pages. Therefore, while a lot of the work of these scholars is solitary, networks have a very important function in sustaining this work, and ICTs are thoroughly enmeshed with this practice. It is not so difficult to dream of ways in which the maintenance and display of network relations might be enhanced via new applications.

Conclusion: Toward a Non-Computational E-Science

We have shown that there is an intriguing tension between the ambition to make e-science the new paradigm of knowledge creation and the actual research practices that are built into the new knowledge infrastructures. E-science, at least as it is embodied in the UK e-science program, is about building a particular type of infrastructure for research. This infrastructure has been shaped in the context of a very specific epistemic culture which originated in the world of high-energy physics and was subsequently modified in computer science and bioinformatics contexts. Nearly all tools that are being built are related to computational research. The problem of whether or not this may constrain the future potential of e-science as an infrastructure conducive to all sorts of academic scholarly work is usually not raised in the e-science community. The spread of e-science is more frequently framed as a problem of awareness raising or of diffusion of ideas and technologies. In other words, there may be a serious problem of misalignment between the emerging e-science community and other scholarly communities. We have conceptualized this as a misalignment between the script that underlies the present massive investments in e-science projects on the one hand and the practices in other epistemic cultures, taking women's studies as an example.

In other words, this tension between the undefined generic ambition of the e-science dream and the particular pilot projects that embody the dream is not an accidental phenomenon but is the very consequence of the way e-science is constructed as a future mode of scientific research. As a result, the *general* description of e-science has an all-encompassing nature. There seems to be no field of research to which e-science might *not* be relevant.

On the other hand, if we look at the *concrete* embodiment of e-science in demonstrator projects, computational research seems to be dominant to the point that it is hardly about anything else than the design of infrastructures for large-scale data and computation. These projects therefore predominantly highlight the Grid as a potential technology for large-scale data management and computational analysis. Another application on which e-science attention is focused is data visualization in modeling and simulation research projects. E-science is therefore strongly associated with computational research (Brockman, 2001; Chien, Foster, & Goddette, 2002; Hey & Trefethen, 2002). The potential for non-computational uses of e-science tools may be mentioned but usually only in passing. In other words, the *promise* is the opening up of venues undreamt of for researchers, irrespective of their specialization. The *practice* is one of building e-science infrastructures by computer scientists in close cooperation with scientists in computational research in physics, life sciences, materials science and social and behavioral sciences.

The various aspects of epistemic cultures analyzed in the section on women's studies can shed light on what a "non-computational e-science" might become. Without wanting to claim that such an analysis would provide guarantees of providing useful tools, such a starting point does highlight other possible directions for e-science. Thinking about epistemic cultures in relation to tools also highlights the decisions that have already been incorporated into current e-science projects.

It is important to note that the three features of epistemic cultures (objects, experiments and relations between units in the field) are interrelated, and that e-science will not simply "answer needs," but also reconfigure practices. If ICTs seem most present in the social/organizational aspect of this epistemic culture, there may be new practices developing that affect objects and their manipulation. To give one final example from women's studies: Members of a network are aiming to pursue a task online, in a distributed manner. The goal is to bind the network, through the creation of a novel object (a database in this case), the manipulation of which will in turn require adjustments to current practices around objects. Such initiatives require expertise, support and tools, which might be elaborated within e-science initiatives.

Finally, if epistemic cultures can speak to e-science, the methods and concepts of this analytic frame must also be questioned. If this approach has the advantage of interrogating the meaning given to technology, it does, however, potentially reify the notion of culture, and may therefore blind us to some of the novel social relations being enacted through technological networks. The setting of e-science may cause us to revisit the notions of epistemic object and technical object, which are key to the notions of laboratory and experiment in the cultural approach to science. These and other issues surrounding epistemic cultures are discussed by Wouters et al. (2007, in preparation).

We have noted above the rise of new analytical puzzles. New methods may also be needed for science and technology scholars who study epistemic cultures in highly-mediated settings. Knorr-Cetina suggests that ethnographic work might be scaled up to study distributed organizations, such as high-energy physics collaboration. E-mail might be used by the analyst, in the same way that physicists also link up to this network (Knorr-Cetina, 1999, p. 23). E-science tools might also be imagined for this kind of research practice (Beaulieu & Park, 2003; Hine, 2002, 2005; Ratto & Beaulieu, 2003; Wouters et al., 2007 in preparation).

We are not arguing that e-science is a promise that will never be realized, and that we are setting out to debunk it. Nor do we want to argue that we are better off looking into the crystal ball and predicting what researchers will need or will want in terms of concrete technologies. Dreams and promises are a necessary part of change. And only a very close collaboration with researchers can hope to have concretely useable tools as an outcome. Rather, we have tried to show in this chapter that the promises of generic tools are not neutral at all, but rather represent the "culture of no culture" (Traweek, 1988) that physicists and computer scientists, as originators of the Grid promise, have been shown to have. Second, we have argued that there is already a strong tendency for e-science to generate a specific kind of application, towards computationally-oriented tools. We have sketched how ICTs are present in fields such as women's studies, which, while neither computationally-oriented nor involved in large scale digitization projects, are still very concerned with digital media and electronic networks. The presence, extension and development of new ICTs may be imagined from an analysis of epistemic cultures, and lead to forms of e-science that have not yet been elaborated—possibly avoiding "top-down" dynamics which are especially likely to rebuff researchers.

In short, if e-science is to be a dream then the question, "Whose dream is it, anyway?" is a pertinent one.

References

Beaulieu, A., & Park, H. W. (2003). The form and the feel: Combining approaches for the study of networks on the internet [Introduction to special issue]. *Journal of Computer Mediated Communication, 4*(8).

Beaulieu, A., Scharnhorst, A., & Wouters, P. (2005, April). Not another case study: Ethnography, formalisation and the scope of science. In *ESRC and ASCoR Workshop*: *Middle Range Theories in Science and Technology Studies,* Amsterdam.

Berman, F., Fox, G., & Hey, T. (2003). The Grid: past, present, future. In T. Hey (Ed.), *Grid computing: making the global infrastructure a reality* (pp. 9-50). Chichester; West-Sussex, UK: John Wiley & Sons.

Boumans, M., & Beaulieu, A. (2004). Objects of objectivity [Introduction to special issue]. *Social Epistemology, 18*(2-3), 105-108.

Brockman, W. S., Neumann, L. J., Palmer, C. L., & Tidline, T. J. (2001). *Scholarly work in the humanities and the evolving information environment.* Washington, DC: Digital Library Federation, Council on Library and Information resources.

Brown, N., & Michael, M. (2003). A sociology of expectations: Retrospecting prospects and prospecting retrospects. *Technology Analysis & Strategic Management, 15*(1), 3-18.

Butterfield, H. (1931/1965). *The Whig interpretation of history.* London: Norton.

CERN (2005). *What is the Grid?* Retrieved June 6, 2005, from http://gridcafe.web.cern.ch/gridcafe/whatisgrid/whatis.html

Chien, A., Foster, I., & Goddette, D. (2002). Grid technologies empowering drug discovery. *Drug Discovery Today, 7*(20), s176-s180.

Griffin, S. (2005, May 15). *Digital content and cyberinfrastructure.* Lecture at the Oxford Internet Institute, Oxford, UK.

Hey, T., & Trefethen, A. E. (2002). The UK e-science core programme and the Grid. *Future Generation Computer Systems, 18*(8), 1017-1031.

Hine, C. (2002). Cyberscience and social boundaries: The implications of laboratory talk on the internet. *Sociological Research Online, 7*(2), U79-U99. Retrieved July 25, 2005, from http://www.socresonline.org.uk/7/2/hine.html

Hine, C. (Ed.). (2005). *Virtual methods: Issues in social research on the internet.* Oxford: Berg.

Knorr-Cetina, K. (1999). *Epistemic cultures: How the sciences make knowledge.* Cambridge, MA: Harvard University Press.

National e-Science Centre (2005). *E-science projects.* Retrieved January 30, 2005, from http://www.nesc.ac.uk/projects/escience_projects.html

Ratto, M., & Beaulieu, A. (2003). Metaphor as method: Can the development of networked technologies be understood using metaphor analysis? *Society for Social Studies of Science* [27th Annual Conference]. Atlanta, Georgia, USA.

Traweek, S. (1988). *Beamtimes and lifetimes: The world of high energy physicists.* Cambridge, MA: Harvard University Press.

Wikipedia (2005). *Whig history.* Retrieved January 30, 2005, from http://en.wikipedia.org/wiki/Whig_history

Women's Studies (2005). *Central goals and research questions.* Retrieved June 6, 2005, from http://www.let.uu.nl/womens_studies/research/goals.php

World Wide Web Consortium (W3C) (1995, May). *About the World Wide Web.* Retrieved July 29, 2005, from http://www.w3.org/hypertext/WWW/WWW/

Wouters, P. (2004). *The virtual knowledge studio for the humanities and social sciences @ the Royal Netherlands Academy of Arts and Sciences.* Amsterdam: Royal Netherlands Academy of Arts and Sciences. Retrieved July 25, 2005, from http://www.virtualknowledgestudio.nl/en/

Wouters, P., Beaulieu, A., Scharnhorst, A., Hellsten, I., Fry, J., Ratto, M., et al. (in preparation, 2007). Mediation, distributed inscriptions and networked knowledge practices: STS meets the challenge of the internet. In J. Wajcman (Ed.), *New handbook of science, technology and society.* Cambridge, MA: MIT Press.

Endnotes

[1]   See for a discussion of the notion of Whig history: Wikipedia (2005).

[2]   We downloaded all documents from the official Web site of the UK e-science program that contained the descriptions of 195 projects (National e-Science Centre, 2005). We also analyzed the disciplinary background of the scientific committees by looking up the personal home pages of the researchers listed as members.

[3]   The Dutch have taken new initiatives which focus on the humanities and qualitative social sciences (Wouters, 2004).

[4]   Business has also discovered e-science as a novel opportunity, of course, but this is strongly focused to data management and computational problems and on selling e-science software and hardware. A discussion of the impact of e-science on business is outside the scope of this article. IBM is particularly active in the promotion of e-science and Grid computing (http://www-1.ibm.com/grid/).

[5]   Each project can only have one goal, but may contribute to more than one field or may contain more than one type of research. Therefore, the total numbers of the three lists differ.

[6]   The complete list is given in the appendix.

[7]   The complete list is given in the appendix.

8   We do not mean to belittle, of course, the study of e-science practices by social scientists from the perspective of the social shaping of technology, social informatics, or science and technology studies. This book is itself a result of these research efforts. Until now, however, insights from social science have not been taken up in conversations underlying the creation of e-science or cyberinfrastucture programs in the UK or elsewhere, and the building of e-research infrastructures. It is the very purpose of our chapter to point to some important factors contributing to this state of affairs.

9   See for a recent treatment of this question Boumans and Beaulieu (2004).

Appendix

Table 4. Project goals of UK e-science program

UK E-SCIENCE PROJECTS	
Topic	**Nr of instances**
grid architecture	14
visualization	10
genomics	9
virtual organization	8
data	6
hospital Grid	6
Grid services	5
astronomy grid	4
biodiversity	
resource sharing	
security	
climate prediction	3
neuroscience	
physics grid	
proteomics	
networks	
ontology	
aircraft	2
archiving storage curation	
biochemical networks	
complex systems	
computational electromagnetics	
data access	
database integration	
desktop environment	
education	
engineering data	
environmental e-science	
high throughput informatics	
information services	
medical images	
middleware	
ocean diagnostics	
peer to peer	
portal	
vessel design	

Appendix (cont.)

Table 4. (cont.)

One instance each:
aging
anatomy
anthropology
authorization
bacteria models
bandwidth
biomolecular simulation
brain atlas
breast cancer
cancer management
cardiovascular
chemical structure
chemicals design
chemistry Grid
complex materials
computing
computational resources
condensed matter
crystallography
debugging
data mining
disease data
earth system
e-science center
e-science experiences
evaluation,
fluid dynamics
geochemistry
Grid hosting

Appendix (cont.)

Table 4. (cont.)

One instance each:
insulin[e] resistance
integrative biology
Java code
lubrication,
mammography
mathematics
medical devices,
medical diagnosis
messaging
microarrays
military grid
mobile resources
molecular informatics
MS.net software
mouse atlas,
multipart jobs
problem solving environme
protein crystallography
publishing
racing yachts
radiation risk
remote microscopy
search engines
semantic Grid
sharing information
simulation
social science
source querying
streaming
SUN grid
system biology
teleconferencing
television, text mining
trust
university grid

Appendix (cont.)

Table 5. Type of research in UK e-science projects

Type of research	Nr of instances
data tool development	62
Middleware	22
Visualisation	20
grid services	19
distributed/shared computing	17
simulation and modelling	16
resource sharing scheduling brokering hosting	15
collaborative tools	15
grid infrastructure	3
network analysis	
ontology	
software demonstration	
text analysis	
centre of excellence	2
course	
fieldwork support	
performance prediction	
security tool development	
standards	
broadcasting	1
browser development	
building of design environments	
decision support	
experience sharing	
image analysis	
incomplete information management	
information retrieval	
multicasting	
patient care remote access	
quality control information	
repository	
selfmanaging systems	
service quality	
surgery	

Chapter IV

Interest in Production:
On the Configuration of Technology-Bearing Labors for Epistemic IT

Katie Vann
Royal Netherlands Academy of Arts and Sciences, The Netherlands

Geoffrey C. Bowker
Santa Clara University, USA

Abstract

The chapter locates the organization of the technology-bearing labor process as an important object of STS/ e-science research. Prospective e-science texts, so central to the pursuit of innovative technologies, construct images of specific technical product outcomes that could justify future investment; such products in turn imply specific labor contributions. To study the production of IT for epistemic practice is to go beyond an inquiry of IT use and design practices, and to consider decisions that get made about how the skill, commitment, performance and product demand of scientists could be coordinated and stabilized. In bringing these considerations to the fore, the chapter presents findings from a study about a particular e-science infrastructure production project—the U.S. National

Computational Science Alliance—at the turn of the 21ˢᵗ century. The chapter illustrates the organizational dynamics in this case that were bound up with the garnering of interest and commitment of scientists who were funded to build interdisciplinary computational media.

Prospective Texts and the Reproductive Passages of Epistemic IT

Contemporary international interest in e-science reflects and maintains a resilient tradition of prospective discursive practice on the part of computing practitioners. To speak only on behalf of the United States, the document recently published by Atkins (2003) is the young one in a series of prospective reports that have been made to the U.S. National Science Foundation in response to its requests for bases on which to assess the direction of its investments in large-scale computing. Indeed, computing advocates in the United States have been making arguments for continued investment in high-performance computational technologies for several decades. Contrary to von Neumann's early vision that only a few computers would be needed across the United States for scientific research, it seems that a need for technical systems for science is insatiable. We might speak here of a resilient *will to produce epistemic IT*. By "epistemic IT" we mean information technologies that are produced for the stated purpose of being used by scientists in their knowledge production efforts[1].

Such reports, cultural media for the expression of epistemic IT's will to produce, are objects of study for those who care about dynamics that constitute the principal zone of research for scholars of science, technology, and society (STS). They articulate problems to which future social activities should be oriented, and propose solutions to which financial resources should be directed. Some scholars of STS have concentrated on understanding how these kinds of texts work (cf. van Lente & Rip, 1998; Brown & Michael, 2002). These scholars speak of the construction of "prospects," "expectation statements," and the power of texts to mobilize "communities of promise" into the present. As Brown has recently specified:

[F]uture-abstractions are put into circulation in the first place to have a 'performative' influence in real time (Michael, 2000). That is, hype is constitutive, it mobilises the future into the present. It is part of the repertoire through which a narrative path or story line is constructed for technologies (Deuten and Rip, 2000). And, as with any narrative or story,

various 'actors' are scripted into the plot and must perform their part if the story is to be successful. Within communities of promise, expectations structure and organize a whole network of mutually binding obligations between innovators, investors, consumers, regulators and so on (van Lente, 1993; 2000). Technological change is therefore a process of constant oscillation between present and future tenses, between present problems and future solutions. (Brown 2003, p. 6)

Prospective texts are thus important "actors" *of* and *for* science, technology and society: They at once perform and illustrate the kinds of anxieties that have become worthy of interest on the part of multiple techno-scientific actors; and they at once perform and illustrate the power of anxiety to draw various interests together in the pursuit of historically specific social projects.

We can speak of a genealogy of such actors, a series of prospective statements—penned in the name of men called Lax (1982), Branscomb (1993), Hayes (1995) and Atkins (2003)—which have performed epistemic IT's will to produce. They are enjoined by the twofold quality of provision and need. And we know that the societal efficacy of such texts stems in part from their capacity to create accounts *of* and therefore provide the legal/institutional conditions *for* the pursuit of new activities that would, or could, make specific possibilities become actual. We want to suggest further that such prospective texts are therefore *reproductive* technologies that are deployed by those with the will to produce. Reproduction, as Marx teaches us, involves both production and the setting up of the conditions for the continuation of production. Marxian theory is not unified on the question of how precisely to characterize the mechanisms of reproduction: Are they best conceived as cultural or economic forces?[2] But what STS inquiries of the dynamics of prospective discourses make apparent is that such texts are cultural- economic[3] actors that occupy a central place in the reproductive passages of IT production, precisely because of the role they play in providing anticipatory accounts of what technologies are needed and how they could be brought to fruition. In virtue of that role, such texts construct *justifications* for financial investment in new IT development efforts: They give epistemic IT a reason to continue. We may speak, then, of prospective discourses as mechanisms that shape and give life to the reproductive passages of epistemic IT's will to produce.

Such centrality compels us to appreciate, or *denaturalize*, the singular quality of the textual genealogy marked above: The specific will to produce that it performs is characterized by the continuous construction and articulation of *new* needs that prospective epistemic IT would meet. When we understand these texts as related prospects over time, in other words, it becomes apparent that the reproductive passages of epistemic IT's will to produce require the proliferation of IT needs that have yet to be met, or of needs that are themselves only a vague possibility—needs that need to be *created*. The specific content of the epistemic possibilities that they name—and the requirements for new production activity that they imply—continue to change—not only because actual needs can be identified but also because non-actual needs can be entertained as potentially real[4]. It is not a trivial point, particularly if one imagines an alternative way of living in the world: What kind of knowledge might we produce with what we now have, and how might we sustain that relation? Think about it...

Although the characteristic of continuous change (improvement?) is commensurable with discourses of progress, to which many techno-science actors are committed, it also carries with it implications that should be a source of anxiety for advocates of e-science practice. The reason is that, although the particular reproductive strategies of epistemic IT advocates have been successful in motivating various actors to sign their name to emerging efforts, they necessarily engage two sites of social practice—consumption and production—on which their success and social legitimacy ultimately depend[5]. Due very precisely to their futuristic and therefore presently fictitious existence in the prospective texts that create the conditions for continued production, these sites of social practice are unstable.

As some chapters of this collection illustrate, important lines of STS research about e-science focus their attention on the contexts of *consumption* through which new IT products are, or might, be given life. Indeed, consumption, or IT use, has flourished as an object of STS research. Its elevated status as an object of contemporary STS research reflects a long-standing preoccupation of STS research more generally—its concern to build social and political accounts of epistemic practices per se. Epistemic practice is *de facto* a site of technology consumption insofar as the means of gathering, representing, manipulating, sharing and witnessing objects of knowledge are inextricable components of knowledge production (Lynch, 1993; Hacking, 1999).

The elevated status of consumption as an object of analysis also reflects a recognition that *successful* development of IT systems will occur only through a dynamic cooperation between context-sensitive agents of development and users themselves. In highlighting processes such as "situated action" and "articulation work," STS scholars have noted that technologies are always incomplete and in the process of being achieved through informal design practices that occur within sites of technology use. If any technical system is going to make it, it will do so only through the ongoing achievements of its users. In other words, technology is always an unstable process in which its design takes shape (see Gerson & Star, 1986; Suchman, 1987). Such a view is attributable not only to phenomenological-sociological insights about the constructedness of human's relations with tools; it also reflects the insight that those who best understand the work that the technology is going to enable, are in the best position to know what form the technology should take[6]. The site of consumption is supposed to be where the *best* design work occurs.

In this respect—and as a kind of research—STS's relevance for epistemic IT also reflects an intuition that STS scholars' particular way of elucidating epistemic practices might elucidate the dynamics through which technologies will be successfully diffused, how and where new markets might be created and stabilized. STS research is thus rendered able to *contribute* to or *enhance* the prospective technological futures that we inquire. And *that* relevance itself strengthens the relevance of consumption as an object of STS inquiry: Consumption emerges as a research object that generates questions, questions that contribute to the reproductive passages of both STS and epistemic IT[7].

Yet there is an analytical lacuna associated with the primacy of consumption as an object of research on STS dynamics: the production process. A feature of the "ongoing achievements" approach to technology innovation is that consumption is a site of design work. Insofar as design work is taken to exhaust the process of production, the distinction between production and consumption becomes blurred. In other words, "production" is subsumed by "design," and "design" is done by "users," or consumers.

However, it is possible to acknowledge the importance of "ongoing (design) achievements" within consumption, and still maintain that the study of production is neither equivalent to nor subsumable by the study of IT consumption and design. For there are many aspects of "production" about which studies of use and design are not typically concerned. "Production" also involves decisions about who should be financed to work to make prospective IT. How should the value of their work be apprehended and compensated? What institutions should invest to pay them? How should their work be ordered and accounted for? In other words, how should the technology-bearing labor process be *waged*[8]?

Contingency theory in organization studies long ago proclaimed that reasonable organizational configurations are variable, and that they have significant consequences for the realize-ability of specific production aims. In other words, there is no one best way to produce everything; rather, how best to go about producing something depends very much on what one is aiming to produce. And yet such a simple insight tends to be obscured in large-scale production for epistemic IT, precisely because it tends to be carried out by and for academic practitioners who carry with them to the process those apparently reasonable organizational configurations through which their own knowledge production practices are routinely carried out. But what if reasonable organizational configurations for producing epistemic IT differ from those that have been honed over centuries for producing academic knowledge? And what if reasonable organizational configurations for production vary in tandem with the specificity of the technologies that are being sought? We may orient to such questions as STS dynamics worthy of study, and try to shed light on the presuppositions and challenges that are associated with epistemic IT's reproductive passages. The persistence of changing promises—inextricably bound with epistemic IT's will to produce— renders such questions salient.

In the next section we will look at a specific example of a shift in the contents of consumption needs as embedded in a prospective text of epistemic IT, and the implications of this shift for the organization of its associated technology-bearing labors.

Interdisciplinary Expansion

A number of e-science efforts associated with the NSF and the Supercomputer Centers in the U.S. have over the years sought the "grand challenge" that might entrain scientists into their networks. And epistemic IT's will to produce has recently set its sights on a site of consumption that could offer it up. In particular, pleas for the need for multi- or interdisciplinary methodologies are being fruitfully synthesized with the prospect of computational innovation. Fruitfully, that is, through their joint articulation in successful proposals for continued investment in high-performance computing. An articulation of this synthesis runs roughly so: Socially important problems require interdisciplinary knowledge practices that can be enabled by high-performance computation; dealing adequately with socially important problems requires continued investment in the production of innovative high-performance computing technologies that can mediate disciplinary difference.

We may speak here of a shift in prospective consumption practices, and of interdisciplinary expansion as a quality of the contemporary reproductive passages of epistemic IT. And while such an expansion has been fruitful for the continuation of production on the part of epistemic IT, it brings with it specific production requirements that throw into relief the contingency of workable organizational configurations. A comparative reading of two prospective texts written to the American National Science Foundation as a means of conveying the status and future funding prospects for supercomputing facilities will throw this into relief.

Lax (1982) expresses a concern about American preeminence in supercomputing technology, which was perceived to be under threat from Japan, West Germany, France and Great Britain. The perceived problems for the U.S. at the time revolved around issues of access and capacity. Responding to the needs of the existing user bases, new research and development were said to be needed. Lots of knowledge producers in specific domains are using supercomputer facilities and their activities contribute to the nation's productivity, which is clearly socially important. But America is threatened by other nations, because of existing limits to access. And there's a capacity problem. So what we need is to concentrate more effort on the production of high-bandwidth networks. More work is needed in computational mathematics, coming up with software and algorithms that will make power use more efficient.

By the time Hayes comes around (1995), there is still a need to enhance bandwidth capacity and access as articulated in Lax; but Hayes has also learned to allude to a new need, IT capacities for multidisciplinary epistemic practice[9]. In Lax we find only a passing reference to interdisciplinarity as a way of characterizing the ARPANET, which provides an example of the direction that a solution to the problem of access should take (cf. Lax, p. 11, n 2). By contrast, Hayes's reference to interdisciplinarity/multidisciplinarity occurs through gestures to the complexity of the problems with which knowledge production should be able to grapple; this should be a problem to which continued investment in supercomputer centers should be directed. As the Hayes document put it, "The task force believes that as the complexity of problems increases, the emphasis will gradually shift to more support of multidisciplinary activities" (Hayes, pp. 12-13). Apart from its passing allusions to practices such as ecological and environmental modeling, however, Hayes allocates little space either to specifying the content of the kinds of technologies that will enable such interdisciplinary activity or to establishing the presence of existing demand on the part of scientists. It is as if Hayes presupposes a link between complexity and interdisciplinarity, and projects that presupposition into the future in which scientists themselves have a stake.

The intervening years that mediate Lax and Hayes give rise to a particular computation-multi/interdisciplinary synthesis that is amenable to such presumptive projection, that is amenable to faith[10]. We locate that synthesis at the turn of the decade in the collaboratory concept put forth by Wulf[11]. And what is important about the collaboratory concept from the standpoint of the question of organizational contingency is that it marks a qualitative leap in the kind of skill configurations that would be required to produce the prospected epistemic IT.

A document spearheaded by Cerf (1993) is a palpable instance—if not the formal progenitor—of this qualitative shift. Collaboratories were center stage in the document, which was modelled after the watershed white paper on "collaboratories" authored by Lederberg and Uncapher (1989); the latter was informed by Wulf's esoteric, unpublished white paper (Wulf, 1988). Now, the collaboratories concept is often alluded to as a technological prospect that could overcome distance and the exigencies of geographical place. In similar terms as Babbage had deployed in the 19th century to celebrate the printing press, computer-mediated collaboration could transform research into a distributed practice done in shared, virtual space. Cerf noted that a "collaboratory provides a technological base specifically created to support interaction among scientists, instruments, and data networked to facilitate research conducted independent of distance" (Cerf, p. 7)[12]. The resonance of Babbage and the Cerf Report on this point clearly rests in the perceived affinities between the printing press and the internet as a medium capable of giving the text limitless mobility[13].

And yet the more significant moment in the Cerf Report is its respecification and extension of the collaboratory prospect in a crucial respect. It is not confined to a focus on technological means with which to overcome geographical distance— a problem which is largely manageable under the conceptual rubric of access capacity as dealt with in Lax, e.g., its allusion to ARPANET as an exemplary "interdisciplinary network." Rather, Cerf focuses on collaborative technology as a means with which to overcome forms of *disciplinary* difference—means that are in no way achievable with the kind of technology scoped in Lax. Cerf does this by synthesizing the logic of Lax's focus (the problematic of processing capacity and access) with a novel problematic: the complexity of the objects of scientific investigation.

From the standpoint of the cultural-economic burdens of reproductive discourse, the logic of the argumentation is ingenious. Cerf refocuses the problematic of computational capacity on two interlocking problems: the volume and the complexity of data that are associated specifically with pressing social issues. Pressing social problems are so scientifically complex that they are intractable to existing disciplinary practices, and their complexity generates volumes of data that are intractable to existing computational means. Prospective computational technologies are required to mediate disciplinary difference; they would enable the interdisciplinary scientist to manage the magnitude of data engendered by a complexity they needed to understand. For example:

More and more scientific problems demand collaboration for their resolution as a consequence of increasing complexity and scale, a growing amount of which reflects the proliferation of fundamentally interdisciplinary problems. The study of global change phenomena illustrates all of these dimensions: it requires the expertise of oceanographers, meteorologists, biologists, chemists, physicists, experts in modelling and simulation, and others from around the world. (Cerf, p. 1; see also p. 5; p. 12)

There is a reference to geographic distance here, but it matters precisely because of the disciplinary difference of those who are far apart from one another. In other words, Cerf suggested that the inter-methodological demands of the world's pressing problems translate into unique technological require-ments that go beyond those which mediate geographical place. Indeed:

Successfully realizing the goals of such programs will almost certainly require a concerted effort to ensure that the electronic infrastructure supports and sustains collaboration across disciplines, whose tools and data types can differ greatly. (Cerf, p. 20—our emphasis)

The last clause is the important detail, because the prospective consumer activity being constructed here in turn requires the production of integrative interdisciplinary media that enable the simultaneous inclusion and transcendence of existing discipline-specific nomenclatural techniques for data representation and manipulation. That is, the interdisciplinary collaboration gestured to under the rubric of complexity is an appeal to a very specific kind of epistemic practice—a specific kind of IT consumption—precisely because of the specificity of the form of *integration* that computational media were scoped to achieve. Overcoming geographical place is some kind of cultural accomplishment, as ARPANET illustrated. But it is a different kind of accomplishment to have overcome disciplinary place as a configuration of singularities that are themselves technical: The *"tools and data types"* that are associated with interdisciplinary collaboration as envisioned by the Cerf report engendered the need to overcome the specificities of *technical place.* It is this form of interdisciplinary work in which Hayes puts its faith.

The epistemic practices and associated IT appealed to in Cerf—and Hayes—instantiate a qualitative transformation in the kind of social mediation that is being proffered, and, in turn, the kind of technical mediations that need to be produced. The singularity of this shift has implications for the organization of production.

Shifts of this sort are important objects of inquiry for STS and for epistemic IT managers, because they involve a mutation in the forms of social contribution that the realization of prospects would require. And such requirements rest uneasily with the futures that imply them, because in appropriating the skills that are required, production incurs a relationship with other organizational forms and the histories through which specific forms of skilled activity have emerged. Translating differences of technical place is not simply a technical barrier; it is a social challenge that must grapple with the histories of actors' skill as emergent properties of their engagement with specific intellectual concerns. Such concerns may be at odds with those that undergird the construction of prospective interdisciplinary projects.

As an illustration of the point we will discuss some findings from a study that we made between 1999 and 2003 of an epistemic IT development project in which infrastructure for integrative interdisciplinary knowledge practices was being pursued under the rubric of the "National Computational Science Alliance" (henceforth, "the Alliance"). In this discussion, we hope to pursue new kinds of dialogues in the e-science/STS exchange. We hope to illustrate ways in which specific labour requirements, linked in this case to the prospect of integrative, interdisciplinary science techniques, bring with them particular organizational challenges.

The Alliance

The Alliance can be positioned in the genealogy and shift discussed in the previous sections. It emerges as a response to the NSF-RFP (nsf9631) issued January 15, 1996, which comes on the heels of the Hayes Report and reflects the prospects that it adumbrates. Indeed, the Alliance may have done more than respond to and reflect Hayes's prospects, for as we learned from a key Alliance architect[14], "there's not like a separate set of folks who have all the ideas in Washington, and then those out in the field who implement them. It's just one set of people out in the field. And so there's this hopelessly, what appears on any rational analysis to be a hopelessly conflicted system, but that's our system, and that's, we're better at it than anybody else in the world."

The organizational approach that the Alliance adopted, with respect to several interdisciplinary domains, was to constitute teams of computer scientists, supercomputer experts and domain scientists from various natural and physical science disciplines, with industry tie-ins as appropriate. These people were ordered organizationally through a division of labour between what were called applications technology teams (ATs) and enabling technologies teams (ETs). In terms of the overall NCSA organization, ATs were composed of domain scientists who were employed at universities but also funded through the partnership, whose efforts would drive information technology development by using prototype Grid software to "attack large-scale problems of science and engineering." ATs would work closely with the "Enabling Technologies Teams" (ETs), composed primarily of university-based computer scientists and in-house labor (variously contracted employees of the NCSA facility), who would "design Grid infrastructure, integrate large-scale management tools, virtual environments, and collaborative technologies within the Grid." (proprietary document, 1998)

The ATs, covering six scientific fields thought by the architects to be particularly relevant to the social problems of the 21st century, were given the task (or opportunity) to contribute to the building of new technologies that would comply with the NCSA Grid initiative vision and thereby respond to the new knowledge-producing practices that were on the horizon for the various knowledge communities from which they were drawn. The reasoning was that ATs would work in partnership with ETs to permit rapid prototyping of new scientific applications, many of which would enable cross-disciplinary collaboration and its associated integration of data types. New tools from the Alliance would be rapidly deployed in the field, such that demand from the field would be met.

At several levels of its technological promise, the Alliance responded to a prospective need to enable scientists from different disciplines to work together in particular ways. Reminiscent of the particular prospective content that we saw above in the discussion of the Cerf Report, NCSA had committed itself to the task of enabling interdisciplinary science formations to work together in ways that would require overcoming the differences in the data types that character-ized disciplinary formations whose integration would provide useful knowledge about socially important problems. For example, a core EHAT[15] objective was described as follows:

A central problem in the physics of the environment is the coupling of systems at multiple scales. Typically, the various scales and phenomena have been studied by different disciplines and sub-disciplines. As under-standing progresses, interest develops in phenomena that couple principles discovered in separate disciplines. For example, the El Nino phenomenon can only be understood by combining ideas from meteorology and ocean-ography, and can only be quantitatively addressed by models including quantitative representations of atmosphere and ocean[16].

In this example, the promise of interdisciplinary knowledge production efforts would yield a more holistic knowledge of complex phenomena, and this in turn would require specific kinds of tools. As suggested in their descriptions, a central task of the group was to produce an integrative technique that could enable scientists to produce a more "realistic" or "holistic" picture of the environment. The team's goal was to contribute applications for the Grid infrastructure, which would include distributed modelling and collaborative visualization tools that could be used in what they called "problem-solving environments." Such environments would be used for prediction, preparedness and management. A feature of the work task of the group stemmed from the insight that although it "is possible to build a coupled model from a blank slate, often this is unrealistically expensive. Therefore, the quantitative models are frequently accomplished by coupling pre-existing models of the relevant sub-systems" (ibid.). Team mem-bers thus were engaged in the necessarily collective task to create a medium through which historically differentiated languages of their respective commu-nities could become integrated—a holistic epistemic-ontological framework that would enable collaborative problem-solving to ensue. Team members spoke of finding a technique—a meta-language, or sets of equations—that would enable them to integrate—they often used the language "gluing"—their respective models together.[17] Such an integrative technique was therefore burdened by the dual challenge of overcoming while recognizing—subsuming[18]—cultural difference. A noteworthy feature of the Alliance effort, in other words, was the production of media that could simultaneous subsume and express the conventions of specific epistemic histories in a way that could be used by all. A technical translation of prior epistemological commitments.

Allusions to possible knowledge-producing arrangements, mediated as such by specific technical accomplishments, are calls for investment in labours that could engender them. This is important, because it shows the intrinsic link between the ascription of value to specific techno-epistemic IT futures on one hand, and the ascription of value to specific technology-bearing labours on the other. With respect to the case under scrutiny here, the specificity of the prospected consumer activity that the product would enable—interdisciplinary computation—engenders specific technology-bearing labour requirements. We want to look at two connected aspects of these requirements here. The first concerns requirements that stem from the need to achieve the technical adequacy of the applications. The second concerns requirements that stem from the need to build a viable consumer base for them. We will ultimately argue that such skill requirements ground a problematic organizational configuration, due to the inter-organizational investment requirements that they bring with them.

First, the technologies alluded to under the rubric of integrative interdisciplinary media must be capable not only of enabling the flow of information about data objects over distance; they must be able to translate across them through what we might call meta-representational objects that transcend the semantic specificities of distinct disciplinary formations that are trying to collaborate to build new kinds of knowledge.

Imagine a telephone system. Each phone has been programmed to work in accordance with a specific language. The phone system is built so that people speaking different languages can speak to each other. So the system has a standard device built inside it that translates each of the different languages being spoken on each phone—a kind of Esperanto that would enable cultural differences to be maintained while precluding no semantic flow from one to the next.

Such a system is implied by allusions to integrative interdisciplinary media. There is the need for what we might call interdisciplinary nomenclatures—higher-order formalizations that can mediate (subsume, translate) the nomenclatural differences that emerge within the interdisciplinary formation as such. For example, "Environmental Hydrology" consists of sub-disciplinary formations that have been organized around ostensibly different forms of the aquatic. Like many other intellectual pursuits formalized in the late 18th century the hydrological sciences have parcelled off sections of natue as objects of inquiry, and sub-specializations—sub-epistemic formations—have been organized around them. In other words, over time, various subdisciplines that can today be drawn together under the rubric of "environmental hydrology" have developed their respective ways of seeing their empirical domains, ways of ordering data, ways of ordering its relations with the natural world.

So, for example, some of the scientists in the group were knowledgeable in the representation and understanding of river flows, others in runoff, others in storms. The group focused on the fact that computational models of hydrologic systems are typically isolated from each other. Atmospheric models describe the physics of clouds and predict rainfall. River models predict flow in stream channels under varying conditions. Runoff models describe surface flow patterns and help to assess the consequences of development. Models of vastly different scales—atmosphere models deal in kilometers whereas river models measure flow in meters—had to be coupled in a way that could both reflect and subsume—simultaneously transcend and maintain—prior epistemic and, crucially, ontological, choices.

For such a technical feat, production requires that builders know all the possible languages that it needs to translate. In other words, the technical device must reflect the specific linguistic conventions of all the users who would use it[19]. Although computer scientists are knowledgeable in the design of a vast array of computational techniques, the objects they are capable of producing are relatively *decontextualized* when considered in relation to the requirements of integrative disciplinary media: Indeed, decontextualized solutions may be the hallmark of their expertise. They are frequently inexperienced with the particular research objects and the questions of particular scientific domains, as well as the nomenclatures and methods that a given discipline has used to construct and manipulate them. So the design and production process geared for domain-specific media requires people who are knowledgeable about the pressing problems and methodological resources of the discipline. And in order to create integrative media that could enable interdisciplinary collaboration in this specific sense, the production process requires that skills that are specific to a variety of epistemic domains be brought together. In such cases, what is needed are collaborative working arrangements between technically savvy practitioners in specific epistemic domains. This was the approach of the group we studied.

But the challenges are not limited to the design of techniques. The successful creation of the product as such entails a process of diffusion in which an ongoing reciprocal relationship of demonstration and consumption takes place. This task resonates with points ascertained in the Cerf report. On one hand, scientists need interdisciplinary technology. On the other hand, the need for interdisciplinary technology will be an *effect* of the demonstration of its usefulness. Not unlike the process engaged by the proverbial pusher on the street, epistemic IT is in this case bound to a reflexive process in which consumption is both cause and consequence of the need it must presuppose[20]. That is why emphases on "education" are so important. They reflect the recognition that establishing viable consumer bases, or "buy-in," requires a specific kind of effort: The product can be established as such only as an aspect of its acceptance through a process of demonstration.

From this standpoint, a domain scientist emerges as a kind of living conduit for the diffusion of technical prospect: He may both represent the methodological norms or languages of his epistemic domain, and garner the consumption investments in a new product on the part of the members of the communities which he or she represents. He or she is both a representative of the ostensible needs of prospective consumers and a broker for and agent of the production agent who, prospectively, exists to serve them. Team members were well aware of this dynamic, and their world provides a living resolution to the proverbial puzzle of the tree that falls in the woods. The dynamic makes it the case that the production process involves not only the collaborative work between different domain scientists, but also between all the communities whose interests they represent and want to secure. And this means that "technology-bearing labour" just expanded beyond the team members who are financed to build the prospected product, to what we might call the (prospective) *communities of concern* from which they come.

Interest Formation

The organizational strategy that NCSA adopted reflects the astute perception on the part of Alliance architects that the production of integrative interdisciplinary media would require a peculiar and historically unprecedented configuration of skills. Team members from the ATs and ETs would work together to permit rapid prototyping of new scientific applications, which would be deployed in the field and therein become involved in an iterative process of re-specification and design until they were usable by various scientific communities.

Environmental hydrology team members were chosen, as everyone involved maintained, because of their high status in particular sub-disciplinary domains ("sub" being a post hoc relational category with respect to "environmental hydrology"). Such a configuration thus fulfilled the twofold skill requirement of domain knowledge and deep entrenchment in the epistemic communities that were, by implication, covered under the overarching rubric of "environmental hydrology."

From the standpoint of the *skills* needed to achieve the prospective integrative media, the Alliance architects crafted an ingenious strategy. The constitution of

ATs was such that the different sub-disciplinary domain experts covered under the rubric of environmental hydrology could collaborate over time toward the production of integrative media that would enable a novel, integrative epistemic practice. The logic here seemed to be that, being able to orient to integrative media as both a mirror and an extension of their existing epistemic practices, the communities of the various sub-disciplinary domains represented by the team members would come to recognize the value of practicing (interdisciplinary) environmental hydrology[21].

However—and in spite of all the cultural emphasis that is placed on the centrality of knowledge to the economy—the problematic reality with which managers must always deal, is that "production" is not reducible to a configuration of forms of knowledge and knowledge-bearing action. Rather it is irreducibly a process through which financial investment is deployed in the procurement and ordering of human contribution to the realization of specific goals. And what we need to focus our attention on here is that, although the distinction between production and consumption is blurry with respect both to design and to the dependency of success on stable consumer choice, there remains a stark distinction between the two sites of activity. Production and consumption need to be distinguished from the standpoint of the mechanisms that secure the interests of their agents, financial investment being one such mechanism. Consumers' interest in a technology is organized through their relations of use with the technology, but technology-bearing labour's interest in the technology is organized by the investments of others in their effort to produce it.

One of the architects of the Alliance characterized the funding strategy he pursued when we asked him to describe the commitment of Alliance members.

[I]f what we figure out is how you could take all these individually funded, peer review driven, follow your nose, individual investigators, and put in place the infrastructure which is missing, which is how do you then harvest the results that emerge from this Darwinian evolution of ideas, and then system integrate them into something that can actually be used by the community, then, after you have figured that out, you have a giant impact on the rest of the community that isn't involved in [the Alliance]. So, purposely, it turns out that I buy people at 10 cents on the dollar (...) Now the reason that's good, is because it's not enough to buy you. Academics are fairly cheap to buy. You know, whole... But, but what I wanted was to buy them cheaply, even cheaper. I just wanted to buy a little bit of 'em. And then, the ones, therefore the ones that will pay attention, there's a lot of overhead to doing the Alliance, the only ones that pay attention are really the believers. Whereas, if I just bought 'em all out, then they'd just take the money, you know, and do their stuff, whether they believe or not. So, I ask a lot more of people than I pay for.

He chided that he'd found a way to get the whole brain by paying for only part of the body; and when we asked him to clarify what he thought he was paying for, he specified: "I am getting them to take a portion of their attention and remove it from the discovery process, and move it to the systemic integration of that new knowledge process."

Now this was a brilliant strategy in that it actually attracted the participation of people with the skills and social relationships that would be necessary to realize the development activity scoped in the name of the Alliance, while off-loading the financial nourishment of those skills and social relationships to other production organizations—university departments, research centers, other grants—by which they were also nurtured. The technological prospect of the Alliance could become an object of joint investment on the part of others.

However, the strategy continued to be plagued by a problematic dynamic. As occasional audits by the NSF found, accounts of future work to be realized by the team as a whole, accounts which had been used as the leverage points for obtaining funding for the teams, were continuously displaced into the future, or always not yet done. Such displacements rendered team members, relative to their stated reason for existing qua funded labour, as having failed. And it was not strictly because they couldn't garner a committed consumer base. As the manager of one of the teams put it, "the hope was that some of this stuff would just take off because it was so useful. But like everything, installing is a non-trivial task, uh, you know, none of these things are push the button and it works. ...) So you have to see a value in doing it. And, we as a team have never been proactive about, you know, communicating to our community what we've been doing, cause we don't have anything to hand them."

Explanations proliferated as to why this should be the case, and they tended to focus on a twofold point that the "deliverables" had not been spelled out clearly enough, and that there were financial reasons for the forthcomingness of deliverables. The manager quoted above put it this way:

Informant: I think what's gonna happen in the next five years is, there's gonna have to be a much clearer definition of what these teams are doing. There's the amount of money that goes into the teams is small compared to the amount of money any group typically has. Now, that's good, because you're leveraging all that other money. On the other hand, it's also harder to get the attention of people you're giving the money to, cause the bulk of their money's coming from elsewhere and they have to meet those objectives. So, you know, different people are developing their own things.... And if you define some new thing to work on it's not so easy to get everybody to,

*well, because they're leveraging with NCSA, or Alliance money, what
they've got real money for.*

A connection drawn—between lack of clarity about what one is contracted to
do and a relationship between different kinds of money; an allusion to "real
money." Readers invested in the successful organization technology-bearing
labour for epistemic IT now have a burning question: How do we get them to keep
paying attention to what we pay them to do, rather than to their own thing? Is
the reality of money, money that holds attention, a function of its quantity? Must
budgets really be increased? Could clarity of definition hold attention in place?

Our own methodological commitment to understand the histories of the technical
commitments that various team members were bringing to the design process,
required us from the beginning to make field studies of the activities with which
team members were engaged (Bowker & Vann, 2000). We may speak of such
engagements as "production territories" rather than "disciplines" or "departments," because of the variety of the sites to which we were taken to as we
attempted to be with the team members in the places where they went back to
when they left the Alliance. Some were scattered offices in university departments, others, research centers taking up a few floors of a university building and
others were institutes in formerly residential properties purchased and renovated
for the purposes of research.

We found these places by creating relationships with the team members and
asking them if we could come and spend some time with them to learn about what
they were up to in their work. And in these places we met students who were
just learning the ropes, post-docs about to solve the crucial equation, secretaries
who had been around since before it got going, and professors who continued to
be in awe of the possibility of prospects they'd outlined years ago. And what held
each of these territories together were not, strictly speaking, the disciplinary
classifications of the people who worked there, but the projects to which its
people were jointly committed: How can we refine *this* modeling system in ways
that are useful to the people who are figuring out what the fishing limits should
be for the bay outside? How can we get *this* visualization system to show cloud
formations in a way that is both convincing but not misleading? How can we
parameterize *this model* in a way that will simulate *that* storm the way it really
happened?

A sense of "this is what we are trying to do here" was the salient flag in each
of the territories, and each seemed to have a specific system or systems to which
the activities of its members were oriented. Team members were, from the
standpoint of their ongoing activities in their respective territories, themselves
producers attempting to achieve users for their products. Although academics
tend to eschew market language of this sort in descriptions of their work, it was

clear from the interactions at their home sites that they were engaged in competitive struggle with other producers to gain users of the systems that they were building—systems in which they had already invested much thought, time and effort. These struggles were the conditions through which their high status had been achieved.

Each of the territories was its own miniature factory, or Alliance, with its own prospective consumers and its own competitors: reproductive passages, everywhere. These activities were themselves collective efforts—sites of people's "own things," things groups *cared about*—that were nourished by a variety of funding sources and that occurred in competitive relations with others who inhabited the territory. Territories stretch across communities that share concerns; they are niche markets organized by resilient interests, resilient reproductive passages. And territories render money "real," not the other way around.

It is possible to view activities that occur in the territories—activities that are one's own thing—as being *different* from those which interest the Alliance architects, different from those which the team members were funded to carry out. However, it is also the case that the value-adding capacities of domain scientists stem not only from their technical know-how, but more importantly from their social relationships as organized by the specific problems with which they are engaged—their lives in their own territories, their own communities of concern. To be a viable representative of the discipline requires not only that one be knowledgeable about the pressing problems and methodological resources of the discipline, but also that one be actively engaged in these issues with other members of the communities who share territorial concerns.

The intriguing point is that these relationships actually problematize the team members' positions as design labour for interdisciplinary technology. For the relationships hold the attention of the team members to problems far less grand than that which the Alliance prospect presumes matters for society. This militates against the diffusion model that has been the *de facto* and unintended consequence of current funding structures and organizational frameworks: Prospective scientists' actual concerns have not become a recognizable object of financial investment.

The funding approach described by the Alliance architect supposes that what is needed for the systematic integration of knowledge is a piece of what is done by actors in what he or she called "the discovery process." The logic could be stated in different language: Continue to work in your local production territories—we need both your skill and your community—but create a very different kind of product and bring it back to the territory to which you are committed. The erroneous assumption is that such territories—and the concerns which give them meaning—can be left and returned to again.

An important lesson may be drawn from this case if we consider the team members' performance in their relation to the Alliance architect's characterization of the funding strategy. The latter seems to be something like this: If you want to buy a scientist even cheaper than it would cost to pay 10 cents for every dollar that contracting his labour actually requires—and this you want to do on the reasoning that otherwise he or she would simply "do their own stuff"—and still get the return on your 10 cents, then you need to put your money on a scientist that really believes in what it is that *you* are trying to accomplish. But what is this belief—so crucial to the identification of labour on which you are willing to risk what you wager—about? The extra 90 cents that the scientist loses in his relationship with you is apparently to be supplemented by his own belief in the importance of the outcome for which his attention is, by you, financed. In turn, you are not requiring that he or she give everything to you in the sense that they would have to give anything else away. But for this to be plausible, the distinction between everything else that he or she could attend to, and the thing to which you are paying him to attend, must be conflated. In other words, what interests you as his financier must be subsumable by what interests him in toto: He or she must take *part* of what he or she is doing and allow it to be subsumed by what you are doing, rather than the other way around.

And as theorists of boundary object logic in STS circles teach us, with respect to social formations in which diverse positions are brought together, what interests you must be stable enough to hold the attention of all of them while yet malleable to serve the respective, prior interest of each. Alliance architects supposed that such stability would attend the chosen team members' belief in its prospects—that signing on to the Alliance as a member was a corollary of one's belief in its aims, and that such a signature is proof that the prospect itself had been taken into the body of which the Alliance only wanted (to finance) a part. But Alliance practices suggest that even in the presence of such belief, indeed, perhaps, precisely because of it, attention was held by the gravity of other, existing concerns[22].

Concluding Remarks

In appropriating the skills that are required to build integrative media, the Alliance inadvertently took on a relationship with other organizational forms, communities whose concerns had cultured specific forms of skill, communities in whom it was not likewise prepared to invest. The skills in which it invested—the skills of individuals—were predicated on actors' engagement with specific intellectual concerns; those concerns were at odds with those embedded in prospective projects. The issue at stake in this case, then, is that one form of technology-bearing labour's interest is organized by financial investment as formalized in the grant, and another form of its interest is organized by the intellectual concerns of the communities with whom laboring subjects have built their skills over time. The disjuncture between these two forms of interest reflects histories from which the future could only imagine to escape.

Brown has extended the study of prospective discourses to an inquiry of retrospective narrativity on the part of techno-scientists, and of the contradictory dynamics of subject formation among techno-scientific actors. A central point concerns the dynamics of discrepancy, between prospective texts and the events to which they purportedly refer. As the authors note, "Simply because the future is represented in a certain way, it does not follow that techno-social arrangements will uniformly concur with the futures idealized for them" (Brown, 2002, p. 5). Brown's fieldwork reveals the centrality of *uncertainty* in techno-scientists' orientations to prospective activities. Although uncertainty certainly characterizes scientists' narrative constructions of the future, it attains a heightened presence in their retrospective accounts of failed events. And what is noteworthy is the way in which the scientists retrospect on their past prospects as being overstated.

Brown reasons that the techno-scientific actor is actually a split or doubled subject, at once scientist oriented to the telling of truths and entrepreneur oriented to the telling of possibilities as objects of financial investment. Only later, in the process of retrospecting the actual feasibility of specific prospects as outlined in the past, does the activity of the "entrepreneur" become an object of the scientist's *realistic* and present assessment of his own past prospective behaviors. In other words, as he or she tells a story of prospective actions in the past, the scientist is in the position to be sceptical with respect to the possibilities that he or she once—qua entrepreneur—prospected as feasible.

There are many lessons that we could take from this research. For the present chapter, and the point we want to highlight with respect to the reproductive passages of epistemic IT, is that there is no need to assume that scepticism on the part of techno-scientific actors qua realists is limited to their retrospective narration. To require retrospection as a frame for the emergence of the sceptical scientist, is to imply inadvertently that there was no entrepreneur/scientist split in the subject as he or she practiced in the past. Indeed, insofar as scepticism is confined to publicly available retrospective accounts, it is likely that retrospective scepticism is a cultural technology deployed in the service of concealing the split itself. Put differently, there is no reason to assume that the scientists' *prospective* accounts of future possibility are equivalent to a true belief on their part that future possibilities are reasonable projections. What matters, instead, is the capacity of the prospective text to perform the scientists' belief in the possibility, regardless of any internal mental state on his part. The prospective text seems to have this capacity—the capacity to perform the techno-scientist's belief—in virtue of its deployment by the scientist, rather than in virtue of any fit between its prospective content and the belief states of the techno-scientist himself. Indeed, we may choose to say that, "Belief does not need to pass through the cognitive apparatus of the "worker," "consumer," or "bureaucrat" because it can be transferred onto a structural or institutional matrix. This external network can then perform the "objectively necessary" function of belief for them" (Flemming & Spicer, 2005, p. 186). To lend one's signature to the realism of a prospective text becomes functionally equivalent to one's belief in the plausibility of a techno-scientific future.

Due to its irreducible dependence on writing, epistemic practice is a site of consumption amenable to epistemic IT's will to produce. Yet such dependence will render epistemic IT's own reproductive passages unstable, for the function—and beauty—of writing, is to make believers of us all. And interdisciplinary science of the sort scoped in funding proposals is a promising standpoint from which to solve the pressing problems of the day (Hackett, 2000). Indeed, part of the seductiveness of interdisciplinary talk is that it is always referring to something bigger and better than anything that could be said to be only a part of what it will claim to become. Certainly, the quest for holism has seductive affinities with Mertonian norms of scientific rigor and completeness. But current practices suggest that, although many would lend their signature to a belief in its preferability, (sub)disciplinary scientists are not compelled to transform the way they do their work so that their results can be integrated with that of other (sub)disciplinary scientists[23]. Put differently, integrative interdisciplinarity media may be beyond the scope of what any community finds, well, interesting.

Acknowledgments

We gratefully acknowledge that work for this chapter has been supported by the United States National Science Foundation (NSF 0094632) and the Royal Netherlands Academy of Arts and Sciences. We also thank Anne Beaulieu, Sisse Finken, Jenny Fry, Christine Hine, and Paul Wouters for helpful and encouraging feedback on an earlier draft.

References

Althusser, L., & Balibar, E. (1998). *Reading capital* (Revised ed.). London: Verso Press.

Amin, A., & Thrift, N. J. (2004). *The Blackwell cultural economy reader.* Oxford: Blackwell.

Atkins, R. (2003). *Revolutionizing science and engineering through cyberinfrastructure.* Report of the National Science Foundation Blue-Ribbon Advisory Panel on Cyberinfrastructure.

Bowker, G. C., & Vann, K. (2000). *Values into infrastructure.* NSF Grant No. 0094632.

Branscomb, L. (1993). NSF Blue Ribbon Panel on High-Performance Computing. Retrieved July 29, 2005, from http://www.nsf.gov/pubs/stis1993/nsb93205/nsb93205.txt

Brown, N. (2003). Hope against hype: Accountability in biopasts, presents and futures. *Science Studies, 16*(2), 3-21.

Brown, N., & Michael, M. (2002). A sociology of expectations: Retrospecting prospects and prospecting retrospects. *Technology Analysis and Strategic Management, 15*(1), 3-18.

Callon, M. (1986). Some elements of a sociology of translation: Domestication of the scallops and the fishermen of Saint Brieuc Bay. In J. Law (Ed.), *Power, action and belief: A new sociology of knowledge?* (pp. 196-233). Routledge: Boston.

Cerf, V. (1993). *National collaboratories: Applying information technology for scientific research.* Washington, DC: National Academy Press.

Deuten, J., & Rip, A. (2000). Narrative infrastructure in product creation processes. *Organization, 7*(1), 69-63.

DuGay, P., & Pryke, M. (Eds.). (2002). *Cultural economy: Cultural analysis and commercial life.* London: Sage.

Finken, S. (2003). Discursive conditions of knowledge production within cooperative design. *Scandinavian Journal of Information Systems, 15*(1), 57-73.

Fleming, P., & Spicer, A. (2005). On how objects believe for us. *Culture and Organization, 11*(3).

Friedman, J. (1976). Marxist theory and systems of total reproduction. *Critique of Anthropology, 7*(2), 3-16.

Gerson, E., & Star, S. L. (1986). Analyzing due process in the workplace. *ACM Transactions on Office Information Systems, 4*(3), 257-270.

Hackett, E. (2000). Interdisciplinary research initiatives at the U.S. National Science Foundation. In P. Weingart & N. Stehr (Eds.), *Practicing interdisciplinarity* (pp. 248-259). Toronto: University of Toronto Press.

Hacking, I. (1999). *The social construction of what?* Cambridge, MA: Harvard University Press.

Hayes, E. (1995). *Report of the task force on the future of the NSF supercomputer centers program.* Washington, DC: National Science Foundation.

Jensen, C. B. (2005, April). *Sorting attachments: On intervention and usefulness in STS and health policy.* Paper presented at the Practices of Assessment and Intervention in Action-Oriented STS Workshop, Amsterdam, The Netherlands.

Lax, P. D. (1982). *Report of the panel on large scale computing in science and engineering.* Paper sponsored by the U.S. Department of Defense and the National Science Foundation, in cooperation with the Department of Energy and the National Aeronautics and Space Administration, Washington, DC.

Lederberg, J., & Uncapher, K. *(1989). Towards a national collaboratory.* Report of an Invitational Workshop at the Rockefeller University, Washington, DC: National Science Foundation, Directorate for Computer and Information Science.

Lynch, M. (1993). *Scientific practice and ordinary action: Ethnomethodology and social studies of science.* Cambridge: Cambridge University Press.

Michael, M. (2000). Futures of the present: From performativity to prehension. In N. Brown, B. Rappert & A. Webster (Eds.), *Contested futures: A sociology of prospective techno- science* (pp. 21-39). Aldershot: Ashgate.

Shapiro, C. & Varian, H. (1998). *Information rules.* Boston: Harvard Business School Press.

Shove, E., & Correljé, A. (2002, November). *Research programmes: Adding value, filling gaps and building networks.* 4th Triple Helix Conference, Lund, Copenhagen.

Star, S. L., & Griesemer, J. R. (1999). Institutional ecology, "translations" and boundary objects: Amateurs and professionals in Berkeley's Museum of Vertebrate Zoology. In M. Biagioli (Ed.), *The science studies reader* (pp. 505-524). New York: Routledge.

Stengers, I. (1997). *Power and invention: Situating science.* Minneapolis: University of Minnesota Press.

Suchman, L. (1987). *Plans and situated actions: The problem of human-machine communication.* Cambridge: Cambridge University Press.

Twidale, M., Nichols, D., & Paice, C. (1997). Browsing as a collaborative process. *Information Processing & Management, 33*(6), 761-83.

Van Lente, H. (1993). *Promising technology—The dynamics of expectations in technological developments.* Enschede: University of Twente.

Van Lente, H. (2000). Forceful futures: From promise to requirement. In N. Brown, B. Rappert, & A. Webster (Eds.), *Contested futures: A sociology of prospective techno-science* (pp. 43-64). Aldershot: Ashgate.

Van Lente, H., & Rip, A. (1998). Expectations in technological developments: An example of prospective structures to be filled in by agency. In C. Disco & B. van der Meulen (Eds.), *Getting new technologies together: Studies in making sociotechnical order* (pp. 203-229). New York: Walter de Gruyter Press.

Wulf, W. (1988). *The national collaboratory—A white paper.* Unpublished manuscript.

Wulf, W. (1995, December). *Keynote address.* ACM/IEEE Supercomputing Conference, San Diego, CA.

Endnotes

[1] Throughout the chapter we will use the term to qualify a formation of various actors who become engaged in the production of such technologies.

[2] For discussions of this issue see Althusser and Balibar, 1970; Friedman, 1976.

[3] For treatments of the locution see Du Gay and Pryke, 2003; Amin and Thrift, 2004.

[4] Marx also showed that the character of reproduction is both variable and contingent, and that it reflects particular social choices about how the will to produce ought to proceed. Epistemic IT's reproductive reliance on the construction of new needs is therefore a particular social choice, and *its* contingency is an object of inquiry worthy of attention.

[5] It bears mention here that scientists are not passive recipients of technologists' will to produce. The epistemic practices in which scientists participate are a particularly amenable site for the reproductive passages of IT labour, because the economies of the sciences are to a significant extent structured through the progressive creation of need for new knowledge-bearing methods. The peculiar competitive relations in which scientists participate are shaped by the implicit notion that preferable knowledge-

producing arrangements are always on the horizon, and that success and failure are decided through the logic of innovation. In other words, epistemic IT's reproductive logic is actually constructed through two voices—one of provision, the other of need. We are a voice given life through the collective articulation of *lack*.

6   STS's contributions to design theory have been developed over the years *in tandem with* other intellectual traditions whose primary objectives political in intent. For a critical inquiry of the discursive politics of one such tradition – Scandinavian cooperative design—see Finken, 2003. Such inquiries are an important instance of STS research that will become increasingly relevant for STS's own reflexive potentials as an intellectual formation.

7   As a privileged research object, then, use/consumption thus mediates a process through which STS sorts its own attachments to epistemic IT. (On "sorting attachments," see Jensen, 2005.)

8   A wage is not only that by which labours are compensated; it is also the name of that which is staked or ventured…that for which one incurs risk or danger.

9   Following the renewal of four of the Centers in 1990, the National Science Board (NSB) asked the director of NSF to appoint a blue ribbon panel— which resulted in the Branscomb Report: From desktop to teraflop: Exploiting the U.S. lead in high-performance computing, was presented to the NSB in October, 1993. Hayes bootstraps findings of Branscomb and by 1995 provides further status and recommendations about the supercomputer centers.

10   By faith we mean belief in something that does not yet exist—it cannot be an object of knowledge, or belief. To be sure, the expression of faith in this sense happens all over the Hayes report. For example, "The Task Force also believes that there will be significant growth in the number of disciplinary and interdisciplinary areas (for example, ecological modelling, and multi-disciplinary design optimization) that will be significantly advanced as computing capabilities advance and as the relevant scientific and engineering communities develop a cadre of knowledgeable users" (Hayes, p. 48).

11   It is interesting to note here that Wulf's 1995 keynote address to the supercomputing conference is devoted to the spectre of falling U.S. budget allocations to computing as a disciplinary activity. Wulf explained to supercomputer scientists that, in such a situation, the most effective survival techniques would be to collaborate with people from other disciplines (cf. Wulf, 1995).

12   Basic components of such a base are electronic libraries, archives and data retrieval techniques (cf. ibid., p. 57).

13   We can see this reliance on the internet example in Cerf's document: "Although technology will never cause the unwilling to collaborate, it can facilitate collaboration among those who are motivated and can also make it more attractive to others. There is evidence that this is happening. One example is the phenomenal growth in the provision and use of services offered through the Internet, the global network spawned by federally funded research into computer-based communications and now used by millions of scientists, engineers, and educators. Through the Internet, researchers access databases, share software and documents, and communicate with colleagues." (See *Executive summary*.)

14   For purposes of anonymity, we will use the word "Alliance architect" to refer to individuals who had significant roles in envisioning and organizing the Alliance effort.

15   Our initial research design was a comparative study of the Environmental Hydrology AT (EHAT) in its effort to develop coupled model applications, with a secondary focus on the Chemical Engineering AT for comparative purposes. We were particularly focused on the ATs, because their work was where we expected the distributed work of domain scientists bringing disciplinary legacy techniques to the infrastructure design process to be most intensive.

16   "Environmental Hydrology Workbench Motivation." Downloaded from a team member's home page. Textual content also appears on grant applications and other proprietary texts (Retrieved, January, 2000).

17   Epistemological realism and ontological holism were intertwined aspects of the group's accounts of its efforts. The philosophical stances seem to be key to a strategy of othering disciplinary work as a practice to be corrected or transcended. The reasoning goes roughly so: Because natural systems are wholes, disciplinary knowledge can represent them only partially, if not only dubiously. Interdisciplinary knowledge is, by comparison, better capable of knowing natural systems, that is, as wholes, or, as they *are*. The form of othering represents a distinctive moment in the history of the sciences as cultures of legitimation. Although discourses of interdisciplinarity apparently invert the cultural logic of reductionism, it shares with the latter a register of positive explanation that could subsume all others (see Stengers, 1997).

18   We use "subsume" here in its conventional sense, as an act of including an instance, an idea, or a category, in an overarching rule. (Latin: *sumo*—to take).

19   In this respect, integrative media problematize the distinction between technology and information content ("the bottle and the wine") that is presupposed in some treatments of the formation of standards in technology markets (see Shapiro & Varian, 1999).

20   Questions of "uptake" (cf. other chapters in this volume) are crucial, then, because they concern both questions of impact on scientific practices and, importantly, the question of whether a "product" can be instantiated as such. To stabilize the consumption of IT is to be a contributor to the process of its production.

21   In this sense, we find an implicit presumption on the part of Alliance architects, that a process of diffusion enabled by "boundary object," or "translation" logic (Star & Griesemer, 1999; Callon, 1986), would obtain on the part of sub-disciplinary communities. A form of practice could emerge that was, overarching as it was vis-à-vis any of their respective existing local interests, bound to become interesting.

22   For a slightly different take on the point, see Shove and Correljé (2002).

23   See Twidale et al. for a discussion of the role of altruism in this process.

Section II

Communication, Disciplinarity, and Collaborative Practice

Chapter V

Embedding Digital Infrastructure in Epistemic Culture

Martina Merz
University of Lausanne & EMPA St. Gallen, Switzerland

Abstract

This chapter introduces the notion of a "disunity of e-science:" It posits that different epistemic cultures privilege different forms of digital infrastructure, integrate them into their practice in historically and culturally specific ways and assign to them distinct functions, meanings and interpretations. Based on an ethnographic case study of theoretical particle physics, the chapter demonstrates how digital infrastructures are firmly embedded and deeply entwined with epistemic practice and culture. The case is made, firstly, by investigating the practice of distributed collaboration and how it is sustained by e-mail-based interaction and, secondly, by analyzing the practice of preprinting and how an electronic preprint archive has turned into a central element of the scientists' culture. In its conclusion, the chapter cautions against techno-deterministic views of how digital infrastructure might align sciences and turn them into a homogenized "e-science."

Introduction: E-Science in Discourse

E-science is about global collaboration in key areas of science and the next generation of infrastructure that will enable it. (John Taylor, n.d.)[1]

In the future, e-Science will refer to the large scale science that will increasingly be carried out through distributed global collaborations enabled by the Internet. Typically, a feature of such collaborative scientific enterprises is that they will require access to very large data collections, very large scale computing resources and high performance visualisation back to the individual user scientists. (Research Councils UK, n.d.)[2]

"E-science" carries different connotations and incorporates different visions of the future development of science, as is exemplified by the above definitions. In particular, programmatic texts seem to suggest that sciences will follow a common trend toward a new form of scientific research and become aligned or homogenized due to the influence and thrust of new information and communication technology infrastructure. This chapter argues that the image of e-science as a coherent endeavor that encompasses and subsumes a wide range of scientific fields is a rhetorical construction that, while it might serve various purposes, does not adequately mirror the complex and multifaceted nature of scientific practice and culture in an era of widespread computerization. Instead, the view will be advocated that different "epistemic cultures" (Knorr-Cetina, 1999) incorporate, configure and co-evolve with a plethora of (existent, novel and imagined) information and communication technologies in a variety of ways.

Different epistemic cultures are situated very differently in the e-science field, be it with respect to the kind of digital infrastructure they privilege, the timeframes in which they promote, adopt or resist to new ICT applications or the epistemic status that they assign to computer-based practices and products (databases, numerical models, etc.). In this article, the notion of a *disunity of e-science* is introduced and promoted to denote a double logic of differentiation. It makes allusion to the concept "disunity of science" (Galison & Stump, 1996) which indicates the growing awareness of STS-researchers that epistemic cultures differ with respect to important dimensions (see also Knorr-Cetina, 1999). "Disunity of e-science," then, refers to the uneven and unequal development of different scientific fields as concerns their adoption and usage of digital infrastructures (for a related argument see Kling & McKim, 2000). Furthermore, the notion also highlights that different forms of computer-based practice may evolve very differently. In accordance, this article wants to caution against attempts to generalize from specific cases in straightforward ways: E-science trends presumably set by pioneering fields will not necessarily be followed up by other sciences, let alone by other areas of society.[3] As constructivist studies of technology have convincingly argued and illustrated (see, for example, Bijker et al., 1987), the pitfalls of linearly extrapolating the development and diffusion of technology into the future are plentiful. To make its argument, this chapter opts for a case-study approach. Based on an ethnographic field study, it aims to demonstrate how digital infrastructures are deeply entwined with the epistemic cultures in which they assume an important role. The selected perspective allows one to bring out the historically- and culturally-situated co-construction of infrastructure and practice, accompanied by a corresponding degree of contingence. The concept of infrastructure employed is loosely aligned with that developed by Star and Ruhleder (1996) who have emphasized that infrastructure is a "fundamentally relational concept" (ibid., 113), to be conceived adequately in relation to organized practices.

To determine what might be an appropriate field for a case study, consider once again the present e-science discourse. According to the definitions that are quoted in a range of programmatic texts, e-science is designated as a project of the future, to do with "the next generation of infrastructure." Yet, one might contend that "global collaboration" has become a characteristic feature of "key areas of science" already throughout the last decades and that digital infrastructure has played an important role to "enable" it. Theoretical particle physics is one of these areas. While it typically requires neither "access to very large data collections" nor "very large scale computing resources," it has been one of the first sciences to incorporate different forms of digital infrastructure into its routine practice. In this sense, theoretical particle physics may be considered a precursor of (what is today imagined as) a future e-science and, as such, provides an interesting case for detailed empirical investigation. This text draws on a "laboratory study" (see Knorr-Cetina, 1994) conducted at the Theory Group[4] of CERN, the European Laboratory for Particle Physics in Geneva, which I began in the early years of the World Wide Web. The extended duration[5] of the study has allowed me to observe how different forms of digital infrastructure for doing science have been introduced, shaped, configured and experimented with—that is, how they have become entrenched in scientific practice and culture in such a way that physicists take them for granted today. Because of its focus on issues of collaboration and communication, this article will not address the digital infrastructures used for computational purposes notwithstanding their significance in particle physics (but see Merz, 1999 & Merz, forthcoming).

Epistemic Culture of
Theoretical Particle Physics

To discuss how digital infrastructure and scientific practice interlink in the case of theoretical particle physics requires in a first step to zoom in on central elements of its epistemic culture. Theoretical particle physics is a *"thinking" science* (Merz & Knorr-Cetina, 1997). Theorists construct models and theories to improve understanding of the fundamental processes of elementary particle production and interaction. Model and theory construction entail a variety of epistemic activities, such as setting up and working out different forms of computations. Work is performed at the desk and in front of the blackboard, instruments are reduced to the pencil and the computer and processing is realized through writing. Desk activities consist mainly in the exploration of realizations of abstract algebraic structures. Elsewhere (ibid.) we have described the theorists' work as a struggle with the "hardness" of a current problem and we have traced the multiple transformations that such struggles undergo when physicists attempt to solve the respective problem, e.g., when they do a particular computation. The interactional dynamics of such struggles appear to consist of a curious oscillation between "straightforward" algorithmic practice and "being stuck," followed by non-algorithmic practice. Non-algorithmic practices are resorted to when following a pre-specified logic of procedure fails, as it continually does. One of the rationales for theorists to collaborate is to jointly cope with the hardness of a problem by diversifying attempts to get across obstacles and by interactively developing new strategies to attack the difficulties.

Theoretical particle physics is a *collaborative* enterprise. Theorists rarely work alone. Collaborations of theoretical particle physicists typically comprise two to four or five physicists who join forces to tackle a research problem. Collaborations may be tied to a specific project and be of short duration only: In this case, a period of a few months will be brought to a close with the preparation of a joint publication. Collaborations may also be continued or renewed for follow-up projects or new projects later. Theorists are promiscuous as concerns the choice of collaboration partners. For the individual physicist, collaborations are no exclusive affair. Typically, the more experienced theorists participate in several collaborations at a time, whether they cooperate with PhD students, with senior scientists or with both at once. In correspondence with the limited duration of individual projects, theorists regularly search for new projects and—this also implies—for new research problems and ideas for how to tackle them, as well as for new collaborators. Their efforts to multiply and intensify contacts with colleagues have to be understood in this context.

The community of theoretical particle physicists is distributed around the world. Dense cooperation and communication networks span national and geographic boundaries. Multiple personal and institutional contacts *interlink* theoretical particle physicists, working groups and institutes across physically distant locations. Occasions for multiplying contact are plentiful throughout the phases of a theorist's career that take him or her to positions in several institutions, cities and (typically) countries. Theorists do not only exploit occasions as they present themselves. In addition, they deliberately create and stage occasions for extending their pool of acquaintances and, thus, of potential future collaborators. For this endeavor, they favor face-to-face situations and the co-presence of their discussion partners. This may be one reason why theoretical physicists are among the most passionate travelers in science. They participate at conferences and schools; they visit research centers and institutes other than their own where they give talks, casually discuss with colleagues and initiate, elaborate or finish up joint research projects. Research centers such as CERN are of central importance in fostering the connectivity of the theoretical particle physics community. They constitute a "marketplace" where project ideas, specific expertise and available resources are negotiated and become assembled into new projects and configurations of researchers. This can be observed at the CERN Theory Group that hosts several hundred physicists from all over Europe and many other countries each year. At one time about 120 to 150 researchers are present, most of whom will stay for two years at most. The high throughput of theorists at research centers contributes to linking up scientists from particle physics groups at universities worldwide. Research centers and conference sites constitute the privileged local contexts for physicists to refresh existing contacts and initiate new ones. They are important crossroads for present and future collaborators and important catalysts of new projects (see Merz, 1998).

Finally, theoretical particle physics is characterized by its highly *competitive* nature. It constitutes an exclusive subculture to which access is restricted. Restriction of access has a structural dimension. Postdoctoral and faculty positions at universities and research centers are scarce compared to the number of aspirants, which incites colleagues and collaborators to compete for positions. In addition, the field's research dynamics are driven by the existence of "hot" topics and problem areas (e.g., recently the area of duality). Successful work on such topics promises an exceptional gain in reputation, which results in an extended number of physicists being attracted to compete on closely-related topics. This suggests that the competitiveness of theoretical particle physics may be considered a defining feature of its culture: The culture is built on beliefs in individual genius and outstanding performance that are not (and, in the physicists' view, should not be) in reach of every physicist.[6] Theoretical particle physicists cultivate a "culture of scarcity" (Krais, 2002).[7] In such a culture, the quest for attaining full membership is never exclusively an individual affair. It simultaneously binds the aspirants into a common project and culture and urges them to compete. It is this double tension that is illustrated by the internet-based "Theoretical Particle Physics Jobs Rumor Mill" and that explains its existence and durability. The Rumor Mill gathers rumors and certified information about open positions and the ranking of candidates for a large number of universities, different countries being served today by dedicated lists. Rumor mills were first started for the job markets in the U.S. and UK, they are now available also for Germany-Switzerland-Austria, Portugal, Greece, etc. The Jobs Rumor Mill symbolizes theoretical particle physics as a highly competitive yet close-knit culture in which the configuration of the community, about what and who is in or out, is a constant topic of conversation and negotiation, with the corresponding information circulating freely at great speed.[8]

When considering the characteristic features of theoretical particle physics, it should come as no surprise that digital infrastructure is widely used. The remainder of this chapter will address the interaction between digital infrastructure and scientific practice by discussing, firstly, distributed collaboration and e-mail-based interaction and, secondly, the practice of preprinting and the importance of electronic preprint archives.

Distributed Collaboration and E-Mail Interaction

In theoretical particle physics *distributed collaborations*, as I will call the cooperative relations that involve physicists working at relatively distant locations, have become frequent. "Distributed collaboration," as employed here, resonates with expressions that are used by e-science proponents. But more importantly, the expression has another connotation in the field of "distributed cognition" (Hutchins, 1995) from which it takes the idea that distribution has not only a physical dimension (different physical locations) but also a social dimension (the distribution of cognitive processes across the members of a social group). The profusion and rise of distributed collaboration in theoretical particle physics may be illustrated by a few numbers: The rate of theory papers published in *Nuclear Physics B* that rely on distributed collaboration has increased from 18% in 1980 to 38% in 1995 and to 50% in 2004.[9] In 2004 59% (1995: 51%) of all collaborative papers (i.e., those with two or more authors) were authored by theorists affiliated with institutes in different towns, states or continents. These numbers raise a first question: What might be the incentives for theoretical physicists to cooperate across considerable physical distances?

Incentives for Distributed Collaboration

The answer involves a combination of factors that are characteristic of the theorists' epistemic culture. To start with, the diffusion of distributed collaboration is related to theoretical particle physics being differentiated into a range of highly specialized sub-domains, each requiring the mastery of a sophisticated set of dedicated conceptual and computational skills. This implies that the sought-for expertise to complement a research project is not necessarily available at a theorist's home institution. She thus has to either search for a collaborator with the required skills at another location or to adapt the planned project accordingly. Even in the case that specializations cluster at specific research institutions and locations—a phenomenon that is observed at certain institutes—distributed collaboration remains frequent. This is related to the high geographic mobility of theorists. A typical career involves a theorist leaving his or her first research institute after obtaining a PhD to take up one, two or even three consecutive postdoctoral positions at different locations, before acquiring a staff position (or leaving academe). Due to typical career patterns, group compositions constantly change with the result that physicists find themselves distanced from their former "face-to-face collaborators" with whom they might want to continue cooperating. Alternatively, as aforementioned, promiscuity is considered to foster original and innovative research and encourages theorists to look for new research partners, which they often do when traveling. Since it tends to be difficult to fully synchronize the presence at an institute and the duration of a project, many collaborators end up being affiliated with different institutes while they work on a common project.

Distributed collaboration is facilitated by the relative independence of theorists from specific locales and their material environments. Most importantly, institutes constitute a social space in which the scientists meet, discuss, initiate projects and present their work and, last but not least, where they earn a living (for a more detailed discussion see Merz, 1998). In contrast to experimental scientists, theorists do not rely on location-specific (material) infrastructure such as the massive research apparatus in an experimental physics lab or the fully equipped workbenches of biologists.[10] Instead, their research infrastructure is either portable or ubiquitous: portable in the sense of the conceptual "apparatus" that they carry in their minds (or the papers that they stow in their briefcases), ubiquitous in the sense of the computers and computer applications that they are provided with in a standardized form at every location. Different internet applications have become part of the ubiquitous infrastructure. Theoretical particle physicists have used them extensively and customized them to their needs since the mid 1980s.

Using E-Mail for Cooperation

While distributed collaboration predates the internet in theoretical particle physics, it has become inconceivable for scientists today to cooperate across distant locations without using e-mail as a means of communication. The sheer amount of distributed collaboration suggests that the practice of e-mail-based cooperation is extremely widespread. This is not synonymous with claiming that e-mail interaction can fully substitute for face-to-face exchanges or will render traveling obsolete. Matters are less unequivocal. Whether a collaboration will rely uniquely on internet-based interaction throughout the duration of a project depends on many factors, such as the degree of acquaintance of researchers among each other, the type of problem that is to be solved, the kind of stumbling blocks that arise along the way and, last but not least, on the existence of alternatives to communicating by e-mail. Theoretical physicists assume a pragmatic attitude. Instead of questioning whether e-mail is appropriate for collaboration, they rather organize cooperative practice in a way as to exploit the advantages of e-mail while trying to work around its perceived disadvantages. For example, they consider e-mail-based interaction to be more suitable for certain phases of a project than others and act accordingly.

For cooperation to be successful and satisfactory for the involved parties, different interactive requirements are to be met in different phases of a project. Consider the phase in which a new project comes into being. To devise a new project, it does not suffice to search for new ideas in the literature. In addition, theorists expose themselves to changing environments and influences and share their experiences and ideas with a variety of colleagues: they "talk physics," as they say. In these situations, "talking physics" has an inherent drive and dynamics toward a new collaborative enterprise. It is directed toward follow-up activities. While not every conversation in which new ideas and potential collaborators are probed leads to a new project, it is expected to constitute a first step. In this case the initial phase of talking physics is succeeded by a phase of "doing physics" (again, a physicists' expression), starting the deskwork stage of the project, which consists of working out elements of the problem in detail. The collaborative work on a project alternates between phases of talking and of doing physics that need not be neatly separated. Talking physics involves the more conceptual elements of the work. It is of help not only when setting up a new problem but also when realigning the aims of the project along the way, setting priorities as concerns the sought-for results and doing all of the interpretative work and contextualization of results that are required to complete a project.

While interaction is crucial in all phases of a project, the interactive requirements in phases of talking physics differ from those in phases of doing physics. Typically theorists have a preference for talking physics in face-to-face situations. They consider e-mail interaction to be more suitable for phases of doing physics in which it is typically sufficient that collaborators keep one another informed about the progress made, each on his or her side, and agree on the tasks to be accomplished next. In contrast, the interactional requirements of talking physics are more demanding. A visitor at CERN maintains that, through e-mail, "you can discuss something which is technical but I am not able to discuss anything which is a little subtle." Another theorist asserts that "you cannot chat over e-mail" and that you remain limited to raise short questions without going into an extensive discussion of controversial issues. This perceived limitation of e-mail is related to its lack of social cues, its restrained interactivity and its priority of simple text-based formats (for example to communicate about formulas, diagrams or other images as physicists do on the blackboard is very time consuming).

To put it differently, the rather purposive e-mail interactions in phases of doing physics open up an "information space" in which statements circulate. In contrast, talking physics requires a "search space," a space in which tinkering, the attempt to deal with ill-defined problems, can unfold in a dialogical manner. Face-to-face situations seem to be more suited for this kind of *distributed tinkering*. Yet, this does not imply that distributed tinkering by way of e-mail interaction does not take place: In fact, a considerable number of theorists do talk physics by way of e-mail on a regular basis.

The case of three physicists provides an illustration. The three have a common history of cooperation; they have successfully co-authored a number of papers before the following exchange took place. The e-mail message that Nick (located at CERN) sent to Patrick (USA) and Alan (Australia) on a Friday later proved to have initiated a new project of the distributed collaboration:[11]

Well, one thing this vacation did for me is to provide me with new energy to start some new project. But what? It seems (several people independently pointed this out to me) that the latest "hot issue" . . . is the construction of free field representations for quantum groups.

In the remainder of his mail, Nick explains what the hot issue is about, provides references to related preprints, asks if either of the two others is knowledgeable about the preprints and concludes: "Real work starts Monday." Both colleagues take up the thread and reply. Alan writes to Nick and Patrick on the following Monday:

I would like to decide over the next few days if we actually can do something on W-brst, or whether we should move on to bigger (?) and better (??) things. I saw Nick's suggestion, although I haven't looked at these papers—no doubt they are all sl(2)? ...

A few hours later Patrick also replies to his two colleagues:

I have printed out the paper of U and comp on BRST for W-gravity. . . . I haven't read what they have in detail, but an extensive use of Mathematica sounds scary. If we want to do anything in this direction we must do it VERY fast. P.

In a message sent to Patrick and Alan on Tuesday, Nick acknowledges the challenge:

We should be able to improve considerably on what has so far been written on W-cohom; let's crack it this week??!! - N

A new project was born, although the collaborators soon realized that the problem could not be "cracked" in a week and required considerably more time and effort. While these and the following 130 mails that Alan, Nick and Patrick exchanged over the next month cannot be analyzed in detail here, the presented e-mails do provide a flavor of the situated and highly contextualized nature of the interaction, which is larded with indexical expressions that draw on several years of close acquaintance and joint membership in the epistemic culture of theoretical particle physics. The correspondence shows how the theorists have customized, re-assembled and re-interpreted the various features of e-mail to produce an infrastructure that satisfies their specific interactional and collaborative needs. The observations guard against an essentialist reading of what e-mail can or cannot do: In fact, the appropriateness of e-mail is revealed in each case *in relation to* the concrete situation and social setting in which it is employed. In the discussed case, for example, the lack of social cues was not conceived as problematic since the three collaborators knew each other very well. It also turned out that the specific problem was prone to a collaborative proceeding, in which dialogic problem-solving proved to be less important than the mutual coordination of actions, a task which could be performed effectively by way of e-mail interaction.

Sustaining Distributed Collaboration

Distributed collaboration is a characteristic feature of an epistemic culture that is connected, collaborative and highly specialized. Digital infrastructure—the facility to interact by e-mail—sustains distributed collaboration, but so do the theorists' traveling practices. In a distributed collaboration, the physicists balance the different modes of interaction and cooperation as well as their perceived advantages and disadvantages according to the specific requirements of the projects and project phases. Being aware that electronically-mediated interaction does not in all situations substitute for face-to-face encounters, theorists meet up with colleagues and collaborators on a regular basis. Electronic exchanges serve as an extension to and a precursor for direct interaction. This observation also leads to the conclusion that traveling does not become obsolete when theorists connect by e-mail. On the contrary, one might even hypothesize that the ease of e-mail interaction encourages theorists to join in collaborations with physically-distant colleagues, which prompts new desires for traveling in the future.

A corollary of this conclusion is that conferences and workshops tied to specific locations and based on the co-presence of participants will not become obsolete either. In contrast to earlier times, conference participants have lost the privilege of exclusive access to brand-new results: Presentations are increasingly transmitted via the internet and the most recent research results are rendered public via the electronic preprint archive (see below). As a result, the raison d'être of conferences has undergone a slight shift, being valued today especially because of the opportunities they offer conference participants to casually discuss with new and old acquaintances. The formalized conference program provides a frame for these interactions. In the face of a community that is connected through myriad electronic interactions, summer centers such as the Aspen Center for Physics or the Benasque Center for Science in the Spanish Pyrenees thrive. They attract physicists with their promise to provide space for the informal exchange of ideas and support them to do research with minimal distraction in a stimulating atmosphere, as is almost poetically illustrated in the mission statement of the Aspen Center:

Here, the essence of the work lies in thought and communication. Often, it takes place on the benches under the trees, in the halls between the offices, on the trails behind the campus or hiking in the surrounding mountains. There are few distractions or responsibilities, few rules or demands. Physicists work at their own speeds and in their own ways: alone or together, at the desk, at the blackboard or in a chair on the lawn. Frequently, a casual, spontaneous discussion gives rise to a new collaboration. (Aspen Center for Physics, n.d.)[12]

Preprinting and the E-Print Archive

The promoters of the ongoing development of digital infrastructure for the pursuit of scientific research typically associate CERN with the World Wide Web and theoretical particle physics more specifically with the success story of its E-Print Archive (see also Bohlin, 2004). Since August 1991, the fully automated E-Print-Archive provides electronic access to theoretical particle physicists' preprints throughout the entire scientific community. The E-Print Archive was first hosted at Los Alamos National Laboratory (LANL) under the name "xxx.lanl.gov E-Print Archive." When Paul Ginsparg, theoretical particle physicist and the server's founder, took up a faculty position at Cornell University in 2001, the E-Print Archive moved with him to Cornell under the new name "arXiv.org E-Print Archive" (http://arxiv.org). Today, the arXiv groups a number of electronic research archives that are dedicated to different sub-domains of theoretical particle physics as well as to other areas of physics and mathematics, nonlinear sciences and recently to computer science and quantitative biology. To illustrate the scale of the endeavor, consider a few submission numbers. Throughout the year 2004, for example, more than 3,000 preprints were posted to "hep-th" ("High Energy Physics—Theory") and more than 4,000 preprints to "hep-ph" ("High Energy Physics—Phenomenology"), as physicists call them in short.

Upon completion, authors electronically submit their article as a "preprint" to the arXiv. At the same time, they submit the paper to a scientific journal for review or to the editors of conference proceedings or the like for inclusion. On the arXiv the preprints are immediately available for download to all interested parties via the World Wide Web interface or by e-mail, different "mirror sites" providing quick access to geographically locations close by. The research papers carry the tag of their submission date and time and are allotted a number (e.g., hep-th/0504013—05 referring to the submission year, 04 to the month and 013 to the order of receipt in the concerned month). Authors may submit a modified version (for example where they have corrected an erroneous passage) with the original number while the earlier version will remain accessible irrespective of its age. The arXiv provides access to the research papers as they are, without an interposed reviewing mechanism. It is considered a success story because of its very high number of submissions and accesses, its universal distribution in theoretical particle physics and its firm integration into the research practice of physics. There is simply no way around the arXiv.

While being praised by some as an individual act of deliberate transformation of the way physics gets done[13], the development of the E-Print Archive and the role it plays in theoretical particle physics today can be interpreted instead as a product of a specific scientific culture and its traditions. This interpretation starts out from the observation that preprinting is a cultural practice that precedes the development of the E-Print Archive in particle physics. "Preprinting" is meant to refer to the practice of both using and producing preprints, preprints being research papers that are made available to an interested public before publication in a journal or a collection of articles, which typically implies before they have undergone quality control by a formalized peer-review process.

Preprints Before the E-Print Archive

In theoretical particle physics, preprints and preprint lists have a long tradition. From the 1950s on, preprint distribution systems relied on the practice that authors (or their institutions) sent a paper copy of their articles to extended distribution lists and a number of close colleagues. The libraries of both CERN and SLAC (Stanford, US) compiled preprint listings on a weekly basis. These lists contained the name of authors and the titles of all preprints that had been received by the respective laboratory (CERN or SLAC) in the preceding week. The lists were sent out to numerous interested parties worldwide. Since this practice was widely known, the two centers assumed the role of a distribution center for information on a vast amount of preprints. Of course, compared to today's electronic archive, the available information was very limited: Whoever was interested in the actual paper had to order it in a separate step from its authors. Ordering preprints had become a standardized procedure that relied on sending a printed order form in postcard format to the authors. The multi-step distribution process was lengthy and physicists had to wait up to several months to receive a recent preprint. Physicists at central locations (e.g., research centers) took advantage of a more direct, and thus quicker, access to the papers.

An attempt to involve computer networks to set up a preprint distribution system was first undertaken in 1989 by the theoretical physicist Joanne Cohn.[14] Cohn started asking colleagues to send her the computer files of their most recent unpublished papers on "matrix models," a topic she was working on herself. She then distributed the files to a list of friends and colleagues. In the course of the next months, the scope of the hand-selected list whose participants became more numerous, due both to Cohn's systematic recruiting strategy and to a snowball effect, raised wide attention. The fact that Cohn was working at the prestigious Princeton Institute of Advanced Study further assisted in promoting the project. Theorists asked to be included in her distribution list and made their papers available. Not only did they realize that their papers would circulate faster over the internet than by way of ordinary mail and traditional preprint distribution systems but they were also eager to receive papers of the leading scientists as soon as they were out. In the summer of 1991—by then the mailing list included about 180 names—Cohn's colleague Paul Ginsparg developed and initiated the first electronic preprint archive that was meant to substitute for the mailing list and was fully automated.

Both Cohn and Ginsparg's activities involved the development of infrastructure for members of their own community, and in both cases colleagues actively cooperated to set up a system that promised—and later proved—to be useful to the community at large. Thus, today's impersonal and fully automated E-Print-

Archive has precursors that rely on the strong connectedness of the theoretical particle physics community and on the centering practices of research centers.

Bulletin Board or Archive?

Today, the arXiv has become part of the infrastructure of theoretical particle physics and is taken for granted. Theorists post their articles to the arXiv, use it to search for related work and download papers in a routine way on a daily basis. The archive is appreciated because it facilitates the search, acquisition and distribution of articles and research results. While the E-Print Archive can be used in a variety of ways, theorists have integrated it into their daily work in forms that correspond to their culture. A first indication may be provided by the fact that, for many years, theorists have referred to the E-Print Archive simply as "the bulletin board" and only recently began to call it "the archives." Ginsparg (1997) has emphasized the difference between an "informal mode of communication" that characterizes Usenet newsgroups or the like and the "formal mode of communication in which each entry is archived and indexed for retrieval at arbitrarily later times" that characterizes the arXiv. In this perspective, the theorists' reference to "the bulletin board" appears to have been imprecise or sloppy. Yet, if one takes the term seriously, one may read it also as an indication that physicists perceive the arXiv as informal and communicative. This interpretation is not to deny that it assumes important archival functions as well.[15] The contention is rather that the significance of infrastructure is insufficiently accounted for when analyzing it merely in terms of function instead of considering also how it is embedded in daily routines and interpreted in the context of cultural preferences. As a result, the question whether the arXiv is *either* archive *or* bulletin board posits a false opposition. For physicists, the arXiv is a bulletin board, for example, in the sense that it is dynamic and exhibits information about the physicists who post their papers and about the ways they go about their work. Most theorists check authors, titles, abstracts and, often, references of new preprints daily. Submissions reveal the time of day an article was posted (was it in the middle of the night, suggesting that a topic is hot and authors feared that competitors might publish first?); resubmissions reveal that mistakes were discovered and corrected; a paper which is withdrawn (but still leaves a trace) hints at the occurrence of major problems. Besides existing entries, there are missing entries, which raise the suspicion that competitors are still struggling with unsurpassable problems. Moreover, the arXiv is a wonderful tool to check with a click how many papers the officemate has published over the last twelve months or over the course of his career. The most recent submissions and what they might reveal are a topic of theorists' informal conversation as are the rumors of the previously mentioned Rumor List. The fact that the arXiv is perceived in such way as a bulletin board (in addition to its perception as an archive), full of explicit and implicit information and content, is a feature of the strong connectedness of the community. At the same time, it contributes to further strengthening this sense of belonging and identity.

Synchronization in a Competitive Culture

In a competitive culture, a physicist needs to be as quick as possible to put a time tag on her paper and a name tag on her results. In case of "priority disputes," the arXiv is today considered a disinterested arbiter because of its feature to record and display the date and time when a preprint was received. In contrast, the peers who review a paper that has been submitted to a journal are not considered equally disinterested. Maxim, a young theorist at CERN, appreciates the arXiv for the immediate visibility that it provides to a paper. He sees this as a safeguard against attempts of referees to hold back an article because they want to delay its publication or take advantage of its content to further their own work. In a competitive culture, the anonymous referees are not trusted—they are primarily viewed as competitors. In this context, the arXiv is conducive to rendering the state of research more transparent throughout the community.

In the eyes of physicists, the increased transparence and immediate visibility of research papers also reinforces the tendency that work is synchronized across research sites. With the electronic preprint archive, earlier paths of preprint distribution collapsed. The distinction between direct access to preprints (e.g., due to co-presence), mediated access (through libraries or the preprint lists of CERN or SLAC) or missing access (when attempts failed) disappeared. As a result, preprints circulate irrespective of geographic location without any distance-induced time delays. Paul, a physicist in Australia, claims that he might not have been able to work on the newest "fads" in his field without the E-Print Archive, through which he is informed about recent work at the same time as his colleagues at CERN, Princeton or elsewhere.

While the unconfined visibility of preprints may indeed have a positive effect on the synchronization of research activities across different institutions and locations, preprints are but one means among others for physicists to inform themselves about recent developments. Paul, for example, is kept up to date also by his collaborators at different research centers who share information not contained in preprints and by his frequent journeys to research centers and conference locations. The claim that location has lost its importance as a determinant for competitive research participation thus needs to be toned down—if not refuted. Physicists without access to the social and economic resources required to travel widely and to draw on an extended network of cooperation and interaction partners at the most important physics centers will have considerably more difficulty to participate in high-profile research on an equal footing. In this respect, it is still an open question whether theorists in countries whose research institutes receive below-average funding can profit from the proclaimed synchronization in the same way as their colleagues from remote but privileged places. The answer is all the more uncertain as the tendency of synchronization is accompanied by a perceived acceleration of research activities. The abridgement of timespans between the production of results and their visibility throughout the community allows (and asks!) for an accelerated time of response to fresh results. Also this feature seems to privilege the scientists at the center of the community who can mobilize resources with more ease to deliver follow-up research output promptly. Culturally, the perceived acceleration goes hand-in-hand with an increased sense of urgency in the theoretical particle physics community which, again, is closely associated with the culture's highly competitive image and nature.

Conclusion: Disunity of E-Science and Infrastructure

The study of theoretical particle physics has illustrated how digital infrastructure is firmly embedded and rooted in epistemic culture by considering two cases: e-mail and the E-Print Archive. Both are deeply entangled with practice, and their meaning for particle theorists cannot be grasped when viewing them as isolated instances of technology. E-mail-supported communication sustains distributed collaboration while the E-Print Archive has become a central element of the physicists' production and utilization of preprints. Both forms of infrastructure are differently embedded in epistemic culture as a review of the empirical findings shows.

Distributed collaboration, which has become prevalent in a culture characterized by a high degree of specialization and considerable geographic mobility, today invariably relies on e-mail supported communication. Theorists accomplish the interactive and cooperative tasks of a distributed project, among others, by customizing and exploiting the specific features of e-mail. Yet, what is at stake is not only that the infrastructure is molded according to the requirements of a specific project; the project's set-up and how the problem is to be solved are as much a topic of consideration as are the project's situational and interactive conditions. In this process, e-mail interaction does not substitute for face-to-face interaction but provides a supplementary form that physicists draw on very differently in different situations. Work is organized by interfacing face-to-face encounters with phases of e-mail interaction. The fact that scientists more easily engage in distributed collaboration today is a feature as much of a traveling culture as of the promise of ubiquitous and effortless e-mail interaction. As a consequence of this dynamic, traditional forms of face-to-face interaction and cooperation are strengthened concurrently with the rise of distributed collaboration.

Preprinting, that is, the production and utilization of preprints, predates the internet. Preprints have served theorists to communicate research results before the formal publications appear. This has not changed with the development of the E-Print Archive. What has changed is the collapse of timespans between the production of preprints and their availability to the community at large and their quasi-universal distribution and immediate accessibility over the internet. Enlaced in an epistemic culture that is characterized by its highly-competitive and connected nature, the current preprinting practice contributes to a perceived sense that research is synchronized and accelerated. In contrast to claims that open and immediate access to preprints translates into a more equitable participation in the most prestigious research endeavors, I hypothesize rather a reconfiguration of the topography of central and peripheral locations.

The notion "disunity of e-science" was coined (see introduction)—*this is the first aspect*—to highlight that different forms of digital infrastructure resonate differently with central elements of an epistemic culture. This explains why they become more or less firmly engrained in a culture, a process in the course of which they may also be considerably transformed. When comparing the use of e-mail for distributed collaboration and work with the E-Print Archive, one first notes that both are fully accepted and universally employed throughout the community of theoretical particle physics. One reason for this might be that they both show a high affinity with characteristic elements of the epistemic culture: For example, e-mail interaction is conducive to a "thinking" and text-based science. Yet, there are also differences. In the case of e-mail usage for collaboration, theorists capitalize both on the relative flexibility of e-mail—it is adaptable to a heterogeneous set of tasks—and on the fact that they do not need to solely rely on e-mail but can use it as a complementary means of interaction. In contrast to e-mail, the E-Print Archive has more closely defined applications which render it less fluid and versatile. On the other hand, the E-Print Archive represents a digital model of a previously established and widely proved system, which was developed by a member of the community with the support of other members for the community at large. In this sense, it does not come as a surprise that it seems to provide something akin to a perfect fit to the preferences of the community and has become an integral element of its culture. An interesting question for future research is what happens when the model is exported to other communities. While the arXiv has extended to a few other domains where it has proved its acceptance by a high amount of submissions (Ginsparg, 1997), it remains to be seen how its use is embedded into the daily practice and the epistemic culture of the concerned domains. At any rate, one may expect that the arXiv does not as smoothly mold indiscriminately with every other epistemic culture (e.g., within the social sciences or humanities). As the discussed cases have shown, the resonance with an epistemic culture does not concern solely the functional dimensions of infrastructure but as much the social and symbolic dimensions that account also for the cultural preferences of scientists. And these may very well turn out to be quite distinct in different epistemic cultures.

The claim of a "disunity of e-science"—*and this is the second aspect*—ensues from the understanding that different epistemic cultures incorporate, configure and co-evolve with different forms of digital infrastructure in different ways. This implies that successful models of digital infrastructure in a certain epistemic culture cannot easily be exported to other cultures while keeping their flavor, functionality, effectiveness and meaning. Part of their success is closely linked to how an infrastructure is specifically embedded. In this sense, the present study makes a case against an overly optimistic or techno-deterministic view of how such infrastructures may "change" the sciences and align them into a homogenized "e-science." The claim of a disunity of e-science and the analysis of how infrastructure is specifically embedded in epistemic culture is thus to be read as a cautionary tale.

Acknowledgments

I thank Christine Hine and Regula Burri for helpful comments, as well as the particle physicists for their advice and hospitality.

References

Bijker, W. E., Hughes, T. P., & Pinch, T. J. (Eds.). (1987). *The social construction of technological systems: New directions in the sociology and history of technology.* Cambridge MA: MIT Press.

Bohlin, I. (2004). Communication regimes in competition: The current transition in scholarly communication seen through the lens of the sociology of technology. *Social Studies of Science, 34*(3), 365-391.

Cornell News (2002). Cornell professor Paul Ginsparg, science communication rebel, named a MacArthur Foundation Fellow. Retrieved March 2, 2005, from http://www.news.cornell.edu/releases/Sept02/Ginsparg-MacArthur.ws.html

Galison, P. L., & Stump, D. J. (Eds.). (1996). *The disunity of science: Boundaries, contexts, and power.* Stanford CA: Stanford University Press.

Ginsparg, P. H. (1997). Electronic research archives for physics. In *The impact of electronic publishing on the academic community.* Portland Press. Retrieved March 2, 2005, from www.portlandpress.com/pp/books/online/tiepac/session1/ch7.htm

Hutchins, E. (1995). *Cognition in the wild.* Cambridge, MA: MIT Press.

Kling, R., & McKim, G. (2000). Not just a matter of time: Field differences and the shaping of electronic media in supporting scientific communication. *Journal of the American Society for Information Science, 51*(14), 1306-1320.

Knorr-Cetina, K. (1994). Laboratory studies: The cultural approach to the study of science. In S. Jasanoff et al. (Eds.), *Handbook of science and technology studies* (pp. 140-166). London: Sage.

Knorr-Cetina, K. (1999). *Epistemic cultures: How the sciences make knowledge.* Cambridge MA: Harvard University Press.

Krais, B. (2002). Academia as a profession and the hierarchy of the sexes: Paths out of research in German universities. *Higher Education Quarterly, 56*(4), 407-418.

Merz, M. (n.d.). Locating the dry lab on the lab map. In J. Lenhard, G. Küppers, & T. Shinn (Eds.), *Simulation. Pragmatic constructions of reality. Sociology of the sciences—A yearbook* (Forthcoming). Dordrecht: Kluwer Academic Publishers.

Merz, M. (1998). "Nobody can force you when you are across the ocean"—Face to face and e-mail exchanges between theoretical physicists. In C. Smith & J. Agar (Eds.), *Making space for science: Territorial themes in the shaping of knowledge* (pp. 313-329). London: Macmillan.

Merz, M. (1999). Multiplex and unfolding: Computer simulation in particle physics. *Science in Context, 12*(2), 293-316.

Merz, M., & Knorr-Cetina, K. (1997). Deconstruction in a "thinking" science: Theoretical physicists at work. *Social Studies of Science, 27,* 73-111.

Star, S. L., & Ruhleder, K. (1996). Steps towards an ecology of infrastructure: Complex problems in design and access for large-scale collaborative systems. *Information Systems Research, 7*(1), 111-134.

Traweek, S. (1988). *Beamtimes and lifetimes: The world of high energy physics.* Cambridge MA: Harvard University Press.

Endnotes

[1]   Definition by John Taylor, Director General of Research Councils, Office of Science and Technology, as quoted on several web sites, for example: National e-Science Centre (n.d.). Defining e-science. Retrieved January 17, 2005, http://www.nesc.ac.uk/nesc/define.html

[2]   Research Councils UK (n.d.). About the UK e-science programme. Retrieved January 20, 2005, from http://www.rcuk.ac.uk/escience/

[3]   Similarly, Wouters and Beaulieu (this volume) observe that e-science proponents shape e-science infrastructure according to their own disciplinary backgrounds. The authors wonder if this might not result in a "serious problem of misalignment" with the infrastructure needs of other communities.

[4]   In the following, "Theory Group" will refer to the organizational unit that is dedicated to theoretical particle physics at CERN. Officially called "Theoretical Studies Division" till the end of 2003, the group was merged with the Experimental Physics Division to form the Department of Physics in 2004 and is referred to informally as "CERN Particle Theory Group" today.

[5]   I first arrived at CERN in the autumn of 1991 and was affiliated with the Theory Group until 1998. Since then, regular contact with a number of theoretical physicists has allowed me to remain informed about recent developments and changes in their everyday practice.

[6]   For the case of experimental particle physics, see Sharon Traweek's account of "Pilgrim's Progress" (Traweek, 1988, ch. 3).

[7]   Beate Krais has coined this expression to characterize the academic system in Germany generally.

[8]   Interestingly, the rumors that circulate in the community indeed concern more often the changing configurations of faculty groups than the private lives of individual physicists.

[9]   *Nuclear Physics B* is one of the major journals in which particle theorists publish. The numbers result from a "quick and dirty" count by hand. "Distributed collaboration" has been operationalized by searching for theory papers whose authors were not all affiliated with institutions in the same town at the time the paper was submitted to the journal.

[10]   It is acknowledged that also experimental scientists have become or are in the process of becoming more independent of specific locations.

[11]   The project in which the physicists' e-mail correspondence addresses is investigated in more detail is Merz and Knorr-Cetina (1997).

12   Aspen Center for Physics (n.d.). Mission. Retrieved April 5, 2005, from http://www.aspenphys.org/brochure/brochure03.html

13   For example, in the announcement that Ginsparg has received an award by the MacArthur Foundation, it says: "Ginsparg has deliberately transformed the way physics gets done" (Cornell News, 2002).

14   Based on private communication with Joanne Cohn (December, 1996).

15   For example, according to Ginsparg (1997), "[O]ver a third of the requests are for papers more than a year old."

Chapter VI

Networks of Objects:
Practical Preconditions for Electronic Communication

Beate Elvebakk
University of Oslo, Norway

Abstract

The chapter is based on interviews with a group of university scientists in the Department of Chemistry, University of Oslo. It seeks to lay bare some of the processes that the chemists carry out in order to transform their "raw materials" into more or less standardized information that can usually be communicated in the forms of e-mails, etc., in a seemingly unproblematic manner. The chapter argues that this work has been a precondition for the success of electronic communication in the sciences, and that information that has not been through such a process is frequently seen as unfit for electronic media. It follows that further introduction of ICTs in the sciences might not prove useful unless this is taken into account. The chapter includes references to problems and literature from the field of Science and Technology studies.

Introduction

The internet has become increasingly important as a means of communication in the sciences. However, in order for net-based communication to work, the information exchanged must already have been made exchangeable, and exchangeable in the right form. In this chapter, I will try to describe some of the processes that precede electronic communication between scientists, and that serve to make this communication possible. The chapter is based on interviews with a group of 25 employees of the Institute of Chemistry at the University of Oslo, and I will attempt to lay bare some of the processes that these chemists carry out that transform their "raw materials" into more or less standardized information that can usually be communicated in electronic form in a seemingly unproblematic manner. The flip side of this is that other types of communication are not as easily translated into electronic form, and that certain types of information are not easily transferred over the internet.

Subject-Object Relations in Theories of Science

One way of framing the last decades' debates around the status of the sciences could be to see them as concerning the relationship between the subjects and objects of science. Whereas the "standard view"[1] of science has assumed the scientists' subjectivity to be utterly irrelevant to scientific findings, newer approaches have suggested that the actions of the scientists are reflected in the scientific product, so that the final product is not "uncontaminated" by external factors, as previously thought.

Schematically, we may say that the "standard," or "classical," view of science presented the object of science as a thing observed, a thing passively discovered by the scientist. The scientific object was there to be found, and the scientist unveiled a preexisting reality. The nature of the object of science was taken for granted, and the subjects were not seen to contribute anything but the discovery. We could say that this view assumes that the subjects and objects are all fairly passive. One discovers, the other is being discovered.

The field "science studies" is in part a reaction against descriptions such as these. David Bloor argued that sociology should not be content to describe "the circumstances surrounding [scientific knowledge's] production" (Bloor, 1976, p. 3), but should aspire to explain the products of scientific investigation—scientific knowledge. Whereas earlier sociologies of science had aspired only to describe the social institutions of science, the sociology of scientific knowledge (SSK) would go further, and say something about the connection between the social processes involved and the knowledge produced. This opposes the standard view of science, as sociology, the study of social interactions, cannot contribute to the understanding of the products of science if the subjects of science do not influence the objects in any way. If sociology is to have explanatory power over the discoveries of science, they cannot be entirely independent of the social interactions related to their discovery. The subjects leave their mark on the objects.

So the emphasis on the scientific objects of the standard view ("The scientific object is not influenced by the discovery process. The subjects 'passively' receive information from the objects.") was challenged by the focus on the scientific subjects in SSK[2] ("The 'truths' about the objects consists in facts about the subjects."). Knorr-Cetina (1981, p. 136) offers a third solution:

Suppose, now, that we begin with neither subject nor object, but with the concept of scientific practice. . . .

This is the approach I have chosen to take in this chapter. Whereas I contend that scientific truths and objects are to some degree "constructed" (that is, constituted through a practical process that might have been otherwise, and without which they would not exist in their present form), I will not pass any judgments as to whether the subjects or the objects have the final say in this process. I will concentrate on the practice as such, and in this practice, the role of digital apparatus for collecting, processing and communicating data is ever increasing. The "data" to which the scientists refer, are simply one version of the object of science, and these data, along with questions, drafts and finished articles are frequently distributed through electronic networks.

In this chapter, I will investigate how the subjects and objects of science, and their reciprocal relations, are changing as objects and relations are increasingly located in electronic media. In so doing, I will make use of the conceptual pair "same/other" to focus on certain aspects of this development. I will try to show how, during the course of scientific practice, the "objects" of science are going through a process where they start out as being "the other" in a radical sense, but are frequently made to be "the same" as the scientist in order for understanding to emerge. Eventually, the scientist must turn the object into a new kind of being that is always "the other" in relation to the scientist, but indefinitely "the same" as itself. Only when this transformation is completed is the object of science ready to be turned into digital information.

The Different Chemistries

Chemistry, according to the *Encyclopaedia Britannica*, is "The science that deals with the properties, composition, and structure of substances (defined as elements and compounds), the transformations that they undergo, and the energy that is released or absorbed during these processes" (*Encyclopaedia Britannica*, 2005). In this sense, chemistry is a unified science, with a shared and defined object of study. However, the same entry goes on to state that "[t]he days are long past when one person could hope to have a detailed knowledge of all areas of chemistry. Those pursuing their interests into scientific areas of chemistry communicate with others who share the same interests" (ibid.). Within the plethora of subdisciplines and other divisions, there exist at least as many ways of relating to the object of study. Consider these two descriptions of their work, proffered by two different PhD students:

If you look into the laboratory over there where there is so much stuff; that's where I work most of the time. So I have lots of test tubes and stuff and I mix things or boil or purify or distil. (PhD student, organic chemistry)

I do most of my research here in the office, but that is a bit untypical, because I am doing theoretical chemistry, so I only use computers. I sit here and I do some programming, and I run some calculations, that is, I try to use the computer for doing chemistry rather than being in the lab mixing compounds. (PhD student, physical chemistry)

For the first of these students, the computer is merely a distraction from his real work, which consists in doing practical stuff in the laboratory. That is, he relates directly to his objects of study as physical substances. For the second, the computer *is* his object of study. He does not relate directly to any reality beyond that. And this will in turn have repercussions for the ways in which ICTs are used for other purposes. Another theoretical chemist told me:

What we do, our normal work, it's by the computer, it's computer stuff, so it's very close to e-mails, in many ways. And it's very easy, you may exchange—when I have written a programme, worked with it for a while, then I take everything I have written, sweep it up and send it over to the people I collaborate with. So we exchange our stuff all the time, and you can't do that with the stuff they work with in the lab. You can't exchange your materials in the same way, so fast and easy. Everything we do, everything we work with, is available and can be sent to the people you work with in ten seconds. (PhD student, physical chemistry)

His work is always potentially exchangeable. It's there to be exchanged. What the collaborators receive is *the same* as what he sees on his own screen. If your work consists in boiling, purifying and distilling, the use of communication media takes on a different role. A process of translation is required for the objects of study to be distributed or exchanged. This chapter seeks to shed light on the differences between the objects that are exchangeable, because they have been made the same as every single copy of themselves, and those that exist—at least for a while—as unique instances.

These two PhD students are extremes. One is doing theoretical chemistry, the other a very practical variety. Most of the chemists engage in relationships with nature that at some point involve a computer, but also an object outside of the computer, which serves as their final point of reference, their anchor in the empirical world. These two extremes, however, may serve to illustrate that the relation to one's object of study is decisive for the kinds of communication one can engage in, and which communication channels may be used.

Inscriptions and Blackboxing: Science as Indirect Knowledge

We should also be aware that even the most hands-on chemist does not necessarily have "direct access" to nature. In Bruno Latour and Steve Woolgar's seminal book, *Laboratory Life. The Construction of a Scientific Fact* (1986), the fictional "observer" tries to make sense of the inner workings of a laboratory through understanding it as a "text-producing" factory. It seems to the naive observer as though not only the entire final output of the laboratory is text in the form of scientific articles, but also as though much of the activity leading up to this can be described in terms of making some kinds of "inscriptions." Material things are constantly being marked, notes are being taken and graphs produced. And whenever notes are available, the material thing has become obsolete, redundant and can be thrown away. The material thing has been replaced by text, and this is what the scientist relates to. The material object—which is avowedly what science is ultimately describing—does not seem to be relevant in its concrete presence, but only as translated into signs. The signs become the object of science; the signs and the objects are considered to be "the same."

The reason why inscriptions are paramount in science is that science deals in abstractions. When dealing with physical objects, sciences like chemistry relate the material thing to conceptual groups that the rest of us do not usually encounter in our day-to-day practices with material objects, even if the material objects in question may be familiar to us. When a chemist works with plastics, which as one of my informants patiently informed me is the lay name for "polymers," he or she does not relate to it in the same ways as we do to our plastic bags.

Chemists have a different kind of practice with polymers. As Ian Hacking points out (Hacking, 1983), scientific practice is a practice not only of representing, but also of intervening in nature. It is a practice of experimenting with atomic structures rather than with carrying plastic bags. The goal of the chemists is to relate the entities in question to broader classes, rather than to describe the individual entity in more detail.

[I]f you try to make it yourself, then—a material can kind of vary from case to case. A polymer is not like sodium chloride that is a substance with well-defined characteristics; polymers vary according to their production. Therefore we depend on working with someone, to work on materials that have been properly characterised, so that we are not only learning something about our own sample, but something more general about the material, or the kind of material that we are studying. (Professor, physical chemistry)

Even though science is supposed to be general, its access to the general must usually go through the particular. The scientist researching polymers, however, would make a fatal mistake if he or she tried to learn as much as possible about the particular sample at hand, before they knew it could be considered representative of a larger class. The individual entity holds no importance in itself, but only as a representative of the class of entities to which it belongs, a class in which all the objects are "the same." The above quote also shows how important it is in scientific activity that the object at hand is well-defined, rather than blurry, and that this may not always be the "natural" state of the object. The scientific object is thus to some degree constructed.

The inscriptions that Latour and Woolgar's naive observer noted are usually not jotted down by the chemist, but are the output of an apparatus. Very much of what is "observed" in a laboratory today, is observed by way of apparatus. Because the use of apparatus is so important to chemical research, chemical articles typically include a paragraph entitled "Experimental," or, more concretely, "Materials and Methods," where the technicalities of the apparatuses (and other materials) employed are listed. An example, taken from an article co-written by a chemist in Oslo begins:

General information. The ¹H NMR spectra were recorded at 300 MHz with a Varian XL-300 (manual) or at 200 MHz with a Varian Gemini 200 instrument.

Varian is the name of a firm specialized in producing chemical instruments, and the XL-300 and Gemini 200 are two of their models. Karin Knorr-Cetina sees this practice of listing the material specifications of the experiment as being a *de facto* acknowledgment of local variation in science, and that "the information obtained in natural or technological science research is idiosyncratic" (Knorr-Cetina, 1981, p. 39). It is certainly an acknowledgment that results are usually not effortlessly "revealed," but require a certain intervening effort.

"Experimental chemistry" is thus very different from what many people imagine. The experiment is more often than not carried out by way of complicated machinery, with names such as *nuclear magnetic resonance spectroscope* or *digital oscilloscope.* Indeed, entire subdisciplines are defined by the apparatus they use for producing results.

In recent years, the apparatus used in experimental chemistry has become increasingly digitalized. A professor, who was working in structural chemistry, described his use of ICTs as follows:

We also use [ICTs] as part of our experimental work. We use PCs—we collect experimental data directly onto the PC, programmes for processing data are on our PCs, and we also use Unix work-stations for processing of data. (Professor, physical chemistry)

Every stage of the experimental process can thus be observed through the machinery, and the machinery is usually connected to a computer so as to make the data easily available and ready for further processing. The "inscriptions" observed by Latour and Woolgar have increasingly become digital inscriptions. Even the inscriptions produced by other instruments are now primarily available through a computer.

In the old days, chemistry was often wet chemistry—you stood there and mixed test tubes and got colours and dregs, but if you look at it today, you have big apparatuses that you feed the samples into, a big apparatus, and then you get it all written by a machine, the results. (Professor, physical chemistry)

These apparatuses, "The so-called material elements of the laboratory," Latour points out, "are based upon the reified outcomes of past controversies which are available in the published literature" (Latour & Woolgar, 1986, p. 85)[3]. The apparatuses are usually the results of earlier findings within this or adjacent fields, and are thus scientific knowledge that has become very concretely "blackboxed": procedures that were once the state of the art of chemistry are now automatically and invisibly performed by these apparatuses. This development has no doubt led to major improvements, and increased the efficiency of chemical practice.

A lot of things go on inside these machines that the researchers could not have carried out by hand, and they represent great scientific progress, but the fact that the researchers have lost some of the control over the process may also feel uncomfortable. A PhD student talked regretfully about the beauty of old scientific work, a beauty that he felt had been lost in today's "automated" chemistry:

[The older chemical writings] are so beautiful it's unbelievable. The results are so accurate; all the uncertainties are always mentioned. Today you have instruments with such high precision, that work very well, but then you have set up your experiments recklessly, and you have these elementary things, so it is not as good as the old work that was carried out with very primitive instruments. So you don't quite get the...sitting there, measuring a point, to sit there and adjust, that's the stuff, right. (PhD student, inorganic chemistry and materials science)

The increasing computerization means that the researchers have less direct access to the object of their own work, it is always mediated, and the mediation processes become ever more complex and opaque as the instruments improve, and incorporate ever more complex processes. The complete knowledge of the procedures involved that was a necessity before the complex apparatus became available is no longer a requirement, and perhaps often practically impossible.

An instrument that is black-boxed is typically standardized. They are bought from an instrument producer who provides the customer with an entire instrumental setup. Evelyn Fox-Keller claims that historically, this standardization of instruments meant that the researcher-subject was made irrelevant. The standardization of the instrument meant that the subject now became the "scientist"

who could speak for everyman but was no-man, in a double sense: not any particular man, and also a site for the not-man within each and every particular observer. Between the seventeenth and nineteenth centuries, a hollow place had been carved out in the mind of every actual or virtual witness into which a machine could vicariously be placed—the lacuna of classical perspective now filled in on the canvas, but emerging, instead, in the mind of the viewer. (Fox-Keller, 1996)

The fact that observations become standardized, because the instruments through which they take place become standardized, means that the subject of science becomes less subject-like. The objects seem to be observing themselves, and a logical line is being established from the object to its description. The objects are no longer subjective descriptions, but objective, and radically other from the scientists, but identical to themselves.

The instruments, then, in addition to helping produce information also help to remove the subject from the object, since the instruments have been manufactured to perform the same procedure, and produce the same outcome in every single instance. The skills and knowledge of the individual scientist have come to matter less once a machine is introduced into the process.

Experiments and Construction

Scientific knowledge is to some degree constructed; crafted through complex practical processes that seem strangely irrelevant once scientific knowledge is established and certain, once it has become "fact." The thing or substance under scrutiny is—during most of this process—available only in a constructed, standardized form, and even before that, it cannot be conceived as an individual entity, but only as a representative of a broader class of beings. Since it is not possible for just any given entity to function as such a representative, the material substances must be carefully selected, and very often constructed in a practical sense of the word. Karin Knorr-Cetina (1981) thus rejects the idea that what is studied in a laboratory is in simply "nature":

All of the source-materials have been specially grown and selectively bred. Most of the substances and chemicals are purified and have been obtained by the industry which serves science or from other laboratories. (Knorr-Cetina, 1981, p. 4)

Similarly, Latour and Woolgar (1986) point out the obvious, yet easily forgotten fact that:

the spectrum produced by a nuclear magnetic resonance (NMR) spectrometer would not exist but for the spectrometer. It is not simply that phenomena depend on certain material instrumentation; rather, the phenomena are thoroughly constituted by the material setting of the laboratory. (Latour & Woolgar, 1986, p. 64)

The "artificiality" or "constructedness" of the nature with which scientists concern themselves is further reinforced by the aforementioned reliance on instruments of inscription. Even the constructed physical objects of the scientific process do not remain the focus of research for a long time. This is a typical procedure:

Usually you can make crystals in one to six weeks. Sometimes even overnight. Then you take one of the crystals and place it in an instrument, and then you get all this [information] that goes into a server, and then you access these data, and process them in a number of different programmes, and with the chemical insight you might possess, and then you may find the structure of this [crystal] and establish that. When that's been done you should study this structure and how this crystal is built, to look at the interaction between the molecules in the crystal, and to what degree this tells you something about your original problem. (Professor Emeritus, physical chemistry)

The crystal in question disappears from view relatively early in the process. It is introduced into a machine, and as the machine collects the relevant information, the crystal itself is made redundant, and the "data" have taken centre stage. The data are now the object of study. As Knorr-Cetina points out, this transformation of a physical entity into a set of data is not a necessary process, but one that consists in a number of selections; certain aspects of the entity have been selected, whereas others are ignored. Thus:

the products of science are contextually specific constructions which bear the mark of the situational contingency and interest structure of the process by which they are generated.[...] Phrased differently, they require that selections be made. Selections, in turn, can only be made on the basis of previous selections: they are based on translations into further selections. (Knorr-Cetina, op. cit., p. 5)

The selections are usually based on previous local or disciplinary history. Scientists choose one project rather than another, one collaborator rather than another, one instrument over another and one method of analysis over the others.

Given that science is conceived as a system in which such selections play an important part, the role of the available instruments and apparatus grows more important. The availability of these instruments is related to past policy-decisions as well as to financial circumstances, but they can also bear a more pronounced local stamp. Thus the object may indeed be shaped by local, contingent circumstances. Certain processes could only take place where the proper instruments were to be found.

We can see that the physical object of science, or "the target," becomes increasingly removed from the subject, not only in principle, but also in practice, as many of the instruments are accessed directly from the PC, which means that all a professor needs to do in order to monitor an experiment is to have a look at his laptop. The physical proximity to the substances is no longer a prerequisite for knowing them intimately from a chemical perspective. However, in this parallel movement, the object—as data—also seems to become increasingly removed from the object—as part of physical nature.

The fact that the "object" is for the most part of the scientific inquiry present and relevant only in the form of measurements means that using the internet for communication is unproblematic: The preliminary results—known as "data" rather than as "facts"—are constantly being sent from computer to computer.

We do some experiments in France, for instance, and then we go down there ourselves and make the measurements, and then we simply send all the data that we lack to ourselves back home in the e-mails. And if we don't manage to bring all our measurements home with us, they are sent to us later over e-mail, and then it works as a mere transfer medium for data, data that somebody measures for us, or we make the measurements ourselves. And then there are students sitting here, doing all the interpretation of the data, and the writing of publications. (Professor, inorganic chemistry and materials science)

The selected aspects of the objects are thus—unlike the original substances— absolutely general and endlessly reproducible and flexible. They can be sent back and forth at will, at no scientific or temporal costs, and with no loss of authenticity involved. The data will be the same, wherever they are. The object, then, has become the same as itself in every single instance or copy, through becoming "the same"—as representation—as the original physical substance, while construed as "the other" in relation to the subjects who partook in their construction.

But more often than sending data to themselves, chemists send data to collaborators, and the data, constructed by machines through measurements and processing procedures, rather than the "object," now serves as the basis for their discussions. The object seems to have all but disappeared and been displaced by "data" that can effortlessly be accessed, transferred, calculated, processed and analyzed in seconds. Or we might alternatively say that the object has been transformed—through what Latour would refer to as a translation—into these data. Or we might say that the data represent the objects, so that they appear to be the same thing. The data have become the same as the object, and thus the object is endlessly identical to itself.

Thus the tendency reported by Latour and Woolgar, where the material substances were replaced by inscriptions, seems to have been reinforced as digital media have become ubiquitous. The inscriptions are more general than ever, as they are accessible from anywhere, and the material aspects of the scientific process have become even more marginalized.

Yes, obviously [the practical work involved in chemistry] is decreasing, nobody here stands by to watch the colours change any more, this basically takes place with machines and sensors, and is logged into a computer.

Q: Except for those who work with synthetic chemistry?

Yes, that's probably the most concrete work, really. But even they usually take their substances to someone else with flashy apparatuses, and try to find whether they have managed to create a pure substance, and so on, so then you're again left with big apparatuses that are connected to computers, and you get graphs that must be interpreted, so really it's all about numbers again. (PhD student, physical chemistry)

In the last instance, there is not actually a question of a physical reference in "nature." The digital apparatus has taken over as the most fundamental reality. There is no reference beyond the apparatus.

We should, however, note that this condition is a reinforcement of a tendency, not a revolution. "Data" were not unknown to scientific practice in the analogue ages. But they were produced, processed and transported in different ways. Before complex apparatus was introduced into the laboratory, data could simply consist in systematic observations, penned down on paper by the attentive scientist. Later, data were also produced by mechanical devices. What is new in the current situation is that data have become potentially omnipresent— extremely malleable and transportable. In this way, data appear to be even farther removed from the physical substances to which they refer. The digitalization of data turns them into a category in their own right, rather than a physical object with a physical history in the laboratory. The digitalization of data thus reinforces data as the same as itself, yet "the other" in relation to any specific physical object, and in relation to any specific person or persons.

The Tacit Dimension

When having discussed the relations to the object in this chapter, we have concentrated on the notion of "inscription." This makes sense of the background to a number of the activities scientists actually engage in, and also of the background to the mostly textual character of electronic communication. But the textual metaphor does not completely convey all aspects of the relations between subjects and objects in chemistry. There is a considerable "tacit dimension" to all scientific activity, and this tacitness can take on different forms and have different reasons, and it may therefore play different roles in relation to ICTs.

Harry Collins (1992, 2001) has established that tacit knowledge is still an important aspect of scientific knowing. Most of the chemists in my study were doing experimental work, in one sense of the word or the other. As we have learned, this is not likely to mean that they work with their bare hands through the entire process, but it does nevertheless mean that there is a practical aspect to their work. This was fully recognized, and perhaps even a source of some professional pride. They would typically say that chemistry was to a large extent based on craft, or skill, and emphasize the importance of hands-on experience, even when what you have your hands on is not the actual thing that you are studying:

It's one thing to explain how you did it, that the others are able to do it, that's another. It's a bit of 'Fingerspitzgefühle' as well. These are small things; you almost have to understand it in order to make it work, too. At least in my field of chemistry, you have to have set it up yourself, and... (PhD student, inorganic chemistry and materials Science)

Thus the feeling of "being one" with the object is seen as important for doing chemistry.

Chemistry has this pre-textual side to it, which the chemists tended to compare to other practical activities, rather than to regular academic work:

You can't read your way into making molecules, or purifying molecules, or analysing them. Because a computer doesn't make molecules. You have to—

practical work is dirty. It's like cooking. You can cook on the screen but it doesn't taste very good. It has very much—practical chemistry, where you make new things—has very much in common with art, and with cooking. And that has to do with—you have to be able to sense, to feel, in addition to your knowledge. You have to be able to see when you mix things together, to sense that if it is about to turn black, then that's an indication that something is going wrong....(Professor, organic chemistry)

Here, again, conceptualizing the object as "the other" becomes problematic. The subject and the object are not always neatly separate entities. Practical work is dirty.

In spite of this awareness of the non-textual skills required in chemistry, nobody seemed to consider this a problem in e-mail communications. One reason that was given for this was simply that collaborators with different areas of responsibility had no interest in what took place prior to the result being ready. But this explanation only holds when the collaborators come from relatively different fields, and collaborate by doing different modules, rather than looking at a problem together. I was told that this holds in about fifty percent of collaborative projects. In the other half, collaborators share more of the same expertise, and may discuss the process more closely. I was nevertheless told that this was never a problem. Typical responses included "we do that in articles all the time," and:

When you work with chemistry, as I do, you make sample, make compounds. And then you may for instance need to know how somebody else made a similar compound, and then I may ask, "what temperature, how did you do that, which were the original substances?" and things like that, and then they may answer, and that works perfectly well over e-mail. Or with some instrument: "you used such-and-such an instrument, and how did you tune it, and so on, what wavelengths did you use for the x-ray-radiation when you did so-and-so?" So these things are...concrete and easy to discuss. (Professor, inorganic chemistry and materials science)

So, the practical work was at once non-problematic and impossible to communicate. There is a transition involved; you go from the private world of physical skill, to the public of standardized measures:

When you have found out how things are supposed to be done, than you can write it down, right, but until you know how to do things, there are elements of...alchemy. (Professor, organic chemistry)

This "alchemist" process is mostly rendered invisible, both by the professional literature, that never reports failures, and by the chemists themselves, who would say that they "simply knew" that some things made no sense to discuss via e-mails. In order to communicate your findings, they must be disentangled from your subject—you must take it from being the same to being the other. This was so much an accepted fact, that no one ever saw it as something that might be improved upon. The chief reason for this is probably that this is the process that makes the chemist into a chemist—the practice of becoming completely immersed in the practical, in order to be able to master it, and come out on the other side as entirely disengaged—a scientist.

Teaching practical skills is only supposed to take place between the students and the supervisor. The others should know. And the person with whom you communicate is understood to have been through the same process:

You can describe this in scientific articles, after all, so the concepts are familiar to those you communicate with.

Q: How do you learn this?

It takes some degree of experience, as well.

Q: But this is not a problem in practice?

No...you trust that the other person has the experience needed for doing whatever needs to be done ... (PhD student, organic chemistry)

Mostly, chemists describe the situation as one of very clear boundaries between the practical skills—which any scientist should already possess—and chemical matters, which can be discussed in a codified form.

Scientific skills were for the most part simply taken for granted, as something that was a necessary requisite of anyone working in the field, something that formed the background upon which scientific discussions were based, but which was never their theme. However, in some research groups, these kinds of areas did not seem to exist at all:

Q: Others have told me that chemistry is often about fingerspitzgefühl?

Yes, but not really in our group.

You don't find it hard to articulate your problems in e-mails?

No, it's so theoretical. (PhD student, organic chemistry)

In theoretical chemistry, the object is already constructed as the "other," but as the same as itself in its various versions and manifestations. Every copy of the file is identical. It does not have to be "forced out" of a recalcitrant material reality.

The chemists see the skills as part of themselves, they must also often tear down the distinction between the knower and the known in this process. They cannot conceptualize the object as the "other" in this process, the tools and instruments they use to find a result must be interiorized, and become part of the scientist. Thus, in the scientific process, the chemists must often go from interiorizing to exteriorizing. They go from making the object the "same" as themselves, to turning it into "the other" to be able to communicate their findings. The distinction between that which cannot be communicated and that which is communicated as a matter of routine often has to do with how far this process has come.

The Importance of Co-Presence

In spite of the perceived usefulness of electronic networks, the chemists insisted that this did not diminish the importance of scientists getting together and being present in the same place. In modern theories of place, the concept has often been approached through the medium of the body (Casey, 1997). Places are the locales where our bodies are, the things and contexts with which our bodies actively engage through acts like walking, grabbing and lifting. In spite of the universality of science, bodily emplacement is nevertheless crucially important for scientific activity, not only accidentally (because all scientists happen to have a body) but also essentially, because science is an embodied type of activity.

Most of the chemists in my study used e-mail on a daily basis, and they saw collaboration and communication via the internet as smooth and unproblematic, but they nevertheless admitted that it was not always an ideal medium for discussions:

E-mail is inconvenient if you have a lot of discussion back and forth. It may be difficult to have a discussion, at least if people haven't got the same experience. In that case it's better to point and show. (PhD student, organic chemistry)

We should take notice of this reference to "the same experience." This suggests that the science is carefully constructed also through the training of scientist. The scientist is part of the scientific structure, and in order for the system to function, the scientists must have become sufficiently similar to be able to communicate (see Haythornthwaite et al. (this volume) for further reflection on this process).

It is also interesting to observe how the breakdown in communication is remedied by "pointing and showing." Pointing and showing refer to bodily movements, rather than just words, and are methods of creating a shared *place*. It seems we can at least partly identify place with what Suchman has termed "situation." For her, "the situation of action can be defined as the full range of resources that the actor has available to convey the significance of his or her own actions, and to interpret the actions of others." (Suchman, 1987, p. 118) This interpretative apparatus is severely limited in electronic communication.

Because the chemists were generally very happy with electronic communication, they could find it difficult to explain why they sometimes preferred, and required, face-to-face meetings. They therefore sometimes imbued spatial proximity with near-mystical qualities:

I think getting together is something else. You can't really replace that; people sitting around a table and having thought the problem through from different angles, and unfolding notes and graphs, and letting the discussion loose. E-mail can never turn into a medium that can replace that. It's a bit impractical, only one person talking at the time. I think perhaps the quality of the results that you communicate by e-mail may be equally good, but the creative thinking is not necessarily that good. Maybe it's necessary for creative thinking that a few different people sit down together around a table and let their thoughts revolve around a problem. I think so. (Professor, inorganic chemistry and materials science)

Here, again, is a reference to using papers and visual aids. In general, three specific reasons were given why bodily co-presence was required:

1. *The speed of communication/response rate.* Many found the rhythm of communication hampered by the strict chronology of e-mail exchanges.
2. *The possibility for showing rather than telling.* Explaining something complicated and frequently half-digested in words was considered to be very difficult, and the availability of a shared visual space alleviates some of these problems.
3. *Interpretation of moods and persons, such as understanding jokes.* Thus, knowing the people with whom you were communicating was seen as an important prerequisite for successful communication by many of the informants. Knowing as person's style, and being able to interpret his or her writings correctly did not require constant, or even frequent, co-location, but it was considered practical to have a certain real-life experience of the person as a background for communication.

In addition to this, it appeared as if the "mystical" aspect of co-location—apart from the sum of the three above factors—often consists in a certain mood, or enthusiasm, that may arise when people come together. It was obvious that when talking about meetings or conference conversations, and similar events, the chemists described an atmosphere of mutual interest and commitment to the situation that they never described when referring to online discussions or e-mail-exchanges. Computer-mediated communication seemed to have a hard time recreating the enchantment of physical co-location.

The lack of visual orientation, however, can be remedied without the actual physical co-presence:

This is something we collaborate on, and then something like this arrives [an e-mail attachment with a schematic drawing of molecules], and then I just send him...these are atomic structures, and if I am to describe this in words, and say "Do you think that one is there?" And now he can just sit over there and I can sit here, and we can send this back and forth, and we may discuss how "This over here is a bit close to that one". This programme that this file is opened in, it can for instance give the distance between this and this [points at figures on the screen], and you can see... [The figures can be moved around on the screen.] And he's got the same programme, and we can use it to discuss chemical relations that we can see. (Professor, inorganic chemistry and materials science)

This quote should also make clear—as in showing, rather than telling—how much easier it is to explain what is going on when two people share the same visual space. The sending of graphical material can to some degree help create the illusion of shared place, and gives increased confidence in the communicative situation.

We use all these drawings, and they can be sent as e-mails, of course. It enables you to collaborate with just about anybody.

Q: You may look at the same thing as you have discussions?

Yes, I'm actually a bit nervous when we can't look at the same thing. (Professor, inorganic chemistry and materials science)

Thus the computer technology can actually contribute to the construction of a more convincing, yet still rudimentary, visual environment. But a mere figure on a screen in itself may not be sufficient:

If you try to extract the essence of your results in eight figures or something like that, you would probably print it out and go walk over and talk to the person in question, and then sit and point with your fingers at the figure. If you want to discuss what's happening and why it's happening, you would probably use figures. An equation is one thing, but there may be a lot of knowledge behind the equation, that you have to express somehow, and then you'll either have to write an awful lot of words, or you'll just have to talk. (PhD student, physical chemistry)

We should note that this quote concerns equations, a mathematical mode of expression. Walsh and Bayma (1997) report that mathematicians use informal communication more extensively than other scientists, reportedly because their highly formalized texts do not properly represent the intuitive and informal nature of the mathematical reasoning process. This claim is supported by Brown (1999), who found that mathematicians reported that they were more dependent on informal communication than scientists from the fields of chemistry, astronomy and physics. Apparently, the relevant features of informal communication are often seen to be missing in communications where the participants are tele-present rather than physically co-present. Words like "creativity" were used to describe what was missing. Again, this might suggest that tele-presence fails to recreate the enchantment of human encounters.

Perhaps what the chemists attempted to describe is not merely co-presence, but what Dreyfus (2001) refers to as *intercorporeality*. Dreyfus holds that physical co-presence is essential, and argues that digital technology, at least as we know it today, will never be able to fully replace the experience of being physically close to another human being. He claims that "our sense of the reality of things and people and our ability to interact effectively with them depend on the way our body works silently in the background" (p. 71f). The silent work of the body provides the deep structure of human interaction: a basic sense of being present, which ultimately implies risk and emplacement in the context. As physical bodies we strive to "get a grip of" the world, an optimum experience of any phenomenon we encounter. This grip is a function of our physical body—we adjust our bodies so as to optimize the grip, we move so as to see or hear more clearly, just as a microscope is adjusted to obtain the optimal view. Since we always *move* our bodies around in order to make the world appear manageable, the static alternatives we are being offered—be they e-mail or videoconferencing systems—cannot provide us with this fuller sense of mastery. Being somewhere is therefore being ready to cope with the world: "Not only is each of us an active body coping with things, but, as embodied, we experience a constant readiness to cope with things in general that goes beyond our readiness to cope with any specific thing"(p. 57). This (physical) readiness to cope with the world Dreyfus defines as a concrete, embodied awareness of *context*.

On the basis of this analysis, he concludes that no distance-communication system, however sophisticated, can ever fully replace the experience of co-presence:

What is lost, then, in telepresence, is the possibility of my controlling my body's movements so as to get a better grip on the world. What is also lost, even in interactive video, is a sense of the context. In teaching, context is the mood of the room. In general, mood governs how people make sense of what they are experiencing. Our body is what enables us to be attuned to the mood. (p. 60)

It seems reasonable to hold that the sense of context is an important part of what is seen to be missing in placeless communication. This context could be anything from the concrete feeling of those present—including their ability to clarify sentences through bodily movements—to the shared background knowledge based on frequent interaction, to the concrete sharing of the same experiences at the same time. Such a sense of context can still be hard to obtain with today's information technologies. The richness of co-location is hard to recreate or replace, even when the problems of communicating over distance have been overcome.

Conclusion

In general, scientists find the internet extremely useful for a number of different purposes, including communicating about their research. In this chapter, I have tried to show that when electronic communication works as well as it does in the sciences, this is partly due to preceding work that renders scientific information "exchangeable" in the relevant way. An important aspect of this process is to sever the ties between the scientist and the scientific results that are being communicated; the data must be rendered autonomous and independent of the individual researcher. Before this process has been completed, electronic communication of research is much more complicated and cumbersome. Usually, however, the work and the skills necessary in this process are taken for granted, and electronic communication is seen as unproblematic.

In spite of this, there are still situations where scientists prefer to meet face-to-face rather than use the internet for discussing their research. This goes for the early phases of a project, where all parameters have not yet been set, and many find it difficult to express their ideas in words, and thus prefer a richer communicational situation. In situations of uncertainty, people also seem to prefer to meet face-to-face, presumably because misunderstandings can easily arise, and because conditions for communication have not yet been settled. Similarly, learning situations are seen to be smoother if they can be based on shared practice, where tacit as well as non-tacit knowledge can be communicated fairly effortlessly.

Part of the perceived shortcomings of present electronic communication channels can probably be remedied, for instance through making use of better graphical tools or real-time communication channels. We should note, however, that part of the attraction of meeting face-to-face seems to be something less tangible, which is probably more difficult to recreate in electronic media. This might suggest that even though the use of electronic communication in the sciences has been successful so far, is not necessarily the case that further use of electronic media will be equally advantageous, as there seem to be significant and relevant differences between the kind of information now seen as easily transferable, and that which is not.

References

Bloor, D. (1991). *Knowledge and social imagery* (2nd ed.). London: Routledge; 1976.

Bloor, D. (1999). Anti-latour. *Studies in the History and Philosophy of Science, 30*(1), 81-112.

Brown, C. M. (1999). Information seeking behavior of scientists in the electronic information age: Astronomers, chemists, mathematicians, and physicists. *Journal of the American Society for Information Science, 50*(10), 929-943.

Casey, E. S. (1997). *The fate of place: A philosophical history*. Berkeley: University of California Press.

Chemistry. (2005). *Encyclopædia Britannica*. Retrieved September 29, 2005, from http://www.britannica.com/eb/article-9022790

Collins, H. M. (1992). *Changing order: Replication and induction in scientific practice*. Chicago: The University of Chicago Press.

Collins, H. M. (2001). Tacit knowledge, trust and the q of sapphire. *Social Studies of Science, 31*(1), 71-85.

Dreyfus, H. L. (2001). *On the internet*. London: Routledge.

Fox-Keller, E. (1996). The dilemma of scientific subjectivity in postvital culture. In P. Galison & D. J. Stump (Eds.), *The disunity of science: Boundaries, contexts and power* (pp. 417-427). Stanford: Stanford University Press.

Hacking, I. (1983). *Representing and intervening: Introductory topics in the philosophy of natural science.* Cambridge: Cambridge University Press.

Knorr-Cetina, K. D. (1981). *The manufacture of knowledge: An essay on the constructivist and contextual nature of science.* Oxford: Pergamon Press.

Latour, B., & Woolgar S. (1986). *Laboratory life: The construction of a scientific fact.* Princeton: Princeton University Press.

Slater, L. B. (2002). Instruments and rules: R. B. Woodward and the tools of twentieth-century organic chemistry. *Studies in History and Philosophy of Science, 33*(1), 1-33.

Suchman, L. (1987). *Plans and situated actions: The problem of human machine communication.* Cambridge: Cambridge University Press.

Walsh, J. P., & Bayma, T. (1997). Computer networks and scientific work. In S. Kiesler (Ed.), *Culture of the internet.* Mahwah NJ: Lawrence Erlbaum Associates.

Endnotes

1   This "standard view" does not correspond to any specific theory of science, as explicit theories are on the whole much more considered views of the scientific process. Some of the points often associated with this view (such as the belief in scientific progress and empiricism) are consistent with various versions of positivism, but as other aspects of positivism—such as anti-realism (Hacking, 1983)—are very much at odds with the standard view, I will not identify the one with the other.

2   There is far from total agreement on the question of subjects and objects within science studies. One fundamental dividing line in the field is to do with exactly this problem. In 1999, David Bloor in an article attacked Latour's theories, and presented the major differences between them in terms of exactly this problem. "Latour's errors about the sociology of knowledge derive from his stance towards a very basic principle which may be called 'the schema of subject and object'. This schema implies that knowledge is to be understood in terms of an interaction between an independent reality, the 'object' of knowledge, and a knowing 'subject', embodying its own principles of receptivity. (Typically, though not neces-

sarily, this subject will be said to construct 'representations' of the object.) Remarkably, Latour wants the sociologist to reject this schema" (Bloor, 1999, p. 82).

[3]   Slater (2002) provides a description of this process in organic chemistry, through focusing on NMR, mass spectroscopy machines etc. He also demonstrates how this "instrumental revolution" in organic chemistry had important repercussions for the understanding of the objects under scrutiny, or "the ontological status of chemical structures."

Chapter VII

Challenges for Research and Practice in Distributed, Interdisciplinary Collaboration

Caroline Haythornthwaite
University of Illinois at Urbana-Champaign, USA

Karen J. Lunsford
University of California, Santa Barbara, USA

Geoffrey C. Bowker
Santa Clara University, USA

Bertram C. Bruce
University of Illinois at Urbana-Champaign, USA

Abstract

As private sector and government research increasingly depends on the use of distributed, interdisciplinary and collaborative teams, particularly in scientific endeavors, we are faced also with an increased need to understand how to work in and study such teams. While much attention has been paid to issues of knowledge transfer, the impact of many other consequences of distribution—disparate disciplines, institutions, career paths, time zones and technologies—have been understudied and underestimated. In this chapter, we describe how distributed, interdisciplinary work puts pressure on existing disciplinary, institutional and personal practices—many of which are second nature to team members, and thus easily overlooked. Reflecting on our own and others' studies of such teams, and our group's experiences as a distributed, interdisciplinary and collaborative unit, we describe some key challenges facing such teams, including issues relating to working and learning together as experts, defining and crossing boundaries, managing external relations and working with and through technologies.

Introduction

Recent trends in work and research have encouraged businesses and research institutions to integrate knowledge from widely disparate fields, to increase the use of computing capabilities and to form inter-organizational connections; in consequence, increasing the dependence on distributed, interdisciplinary and collaborative teams. Private sector investment in alliances has been paralleled by large-scale government investment in research infrastructures, programs and centers, and both have called on researchers to work across knowledge domains, organizational norms and physical and conceptual boundaries. While earlier research has suggested that the main problem for such work is making tacit knowledge explicit for transfer to others, we suggest that contemporary teams face a more complex set of issues as they engage in joint knowledge construction. Contemporary team members find that they cannot simply transfer their previous collaborative skills to a widely distributed, interdisciplinary arena, but must continually renegotiate a wide range of research and work practices thought to be already established. As Knorr-Cetina (1999) has remarked about scientific teams, current research initiatives often bring together quite disparate disciplines, locations and technologies, often leaving a single researcher of such teams woefully at a disadvantage to understand a team's work. Studying contemporary teams requires a more comprehensive examination than is commonly employed, encompassing interdisciplinary processes, group interaction, institutional practices, career interests and uses of information and communication technology. Moreover, contemporary views that consider technology as providing the solution to the "problem" of collaboration—e.g., through faster connection, seamless integration of geographically distributed people and projects and new information and communication technology infrastructures—fail to acknowledge the negotiation of practices and the coevolution of practices and technology that are involved. Collaborations involve dealing with existing embedded practices, as well as emergent ones that take time and effort to evolve.

In this chapter, we draw on previous work in these areas, and, equally important, on the collected experiences of our group—the Distributed Knowledge Research Collaborative (DKRC)—as we observed, studied and participated in distributed, interdisciplinary and collaborative work, to articulate a series of key challenges we find facing contemporary teams. DKRC began in 1997 with the seven future co-PIs composing a grant application to the NSF Knowledge and Distributed Intelligence (KDI) initiative. At the time, all members were co-located at the University of Illinois at Urbana-Champaign: One worked for NCSA (National Center for Supercomputing Applications), and two each in the faculties of Education, Management and Library and Information Science. Members were and remain interested in the impact of technology and new organizational structures on contemporary scientific and work practice. By the time we received funding in 1999, we were located at four universities, and later at five locations across the United States. Starting in 1999, the team has included a number of graduate students who are also distributed across disciplines; an average of about one student per co-PI per year, but with more than seven over the years of the grant.

We have been exploring the ways in which distributed, interdisciplinary and collaborative teams accomplish their work. The teams we have studied include science teams in the fields of biology, cosmology, environmental hydrology, and nanotechnology, as well as social science teams. These teams were chosen as examples of successful endeavors, approved and funded by significant granting agencies, and supported by university and research center infrastructures. We began by interviewing key informants in each team to hear about group organization, administrative procedures and research concerns. We then studied their group collaborative practices using ethnographic, bibliometric, qualitative, and social network methods, and also examined group technologies (e.g., a modeling and visualization tool, digital library and database), and artifacts (e.g., documents, web sites, reports, and research articles; Porac, Wade, Fischer, Brown, Kanfer & Bowker, forthcoming). This work has been backed with considerable discussion and inquiry into what it means to be collaborative in today's scientific world, and how the literature has dealt with the notions of collaboration, knowledge co-construction and the use of technology (see Kanfer, Haythornthwaite, Bowker, Bruce, Burbules, Porac, & Wade, 2000). Throughout our work, we have been deliberately conscious of our own activities as a distributed interdisciplinary team, and how these compare and contrast with those of other groups.

We draw on the following sources in identifying the key challenges outlined in this chapter: the experiences of groups we have studied; our own experiences studying them and operating as a distributed, interdisciplinary team; and the literature on interdisciplinarity, learning and group processes. There are many challenges to any kind of team endeavor, but we concentrate here on character- istics and considerations which have special importance to distributed, interdis- ciplinary and collaborative work groups. We collect the challenges together under these headings:

- Bridging Practices
- Seeing and Crossing Boundaries
- Managing External Relations
- Working with and through Technologies

We do not consider the discussion in this chapter to be exhaustive, but as identifying key factors. These are presented to inform and prepare those embarking on distributed, interdisciplinary work and as a basis for future investigations.

Bridging Practices

Collaborative, interdisciplinary teams are typically composed of experts, or individuals who are highly knowledgeable and proficient in their own work and its practices. Through education, work experience and years of research, teaching, managing and/or mentoring, expert team members have become entrenched to differing degrees within a disciplinary framework. To an expert, disciplinary, institutional and personal research practices are deeply ingrained and often invisible. The challenges experts face in communicating, learning from each other and collaborating are markedly different from those of novices. Where transfer is often the main metaphor for novices, the watchwords for expert interactions are joint problem-solving, shared cognition and co-construc- tion of meaning.

As other writers have explained, experts learn and make sense of problems differently from novices. In contrast to novices, experts approach problems with attention to the principles that can be drawn on to solve them rather than to surface attributes. They are more adept at perceiving patterns, and at under- standing the conditions under which the knowledge will apply. They are skilled at metacognition, "the ability to monitor one's own current level of understanding and decide when it is not adequate" (Bransford, Brown, & Cocking, 1999, p. 47).

In principle (although not always in practice), experts maintain an openness about their state of knowledge and need to learn, and maintain flexibility in their goals (Bransford et al., 1999; Scardamalia & Bereiter, 1996). Yet expertise is firmly rooted in disciplinary and institutional homes. Tacit intradisciplinary understandings about how to examine a problem, what knowledge to bring to bear and how to go about work can get in the way of seamless interdisciplinary collaboration.

Perhaps the most overlooked aspect of interdisciplinary teams is how the embedded *practices* of each discipline form as important a bond as any other part of its domain knowledge. Disciplinary expertise encompasses more than just the facts about the field; it includes knowing what is important for the field, where new problems are emerging, what problems are considered worth tackling and how to formulate and publish ideas and accomplishments to meet grant and promotion requirements. More specifically, it includes field-specific standards in what constitutes data, how they should be collected and shared, what analysis techniques are accepted and when and where data and results are published (see also Fienberg, Martin, & Straf, 1985). Interdisciplinary teams are therefore challenged to integrate not only disciplinary knowledge, but disciplinary practices, including coming to joint understandings about data collection, data sharing, methods and analysis techniques and publishing practices. Such integration can be slow and difficult, in part because we are so entrenched in our ways of approaching problems, and partly because we enter such collaborations without considering integration work to be part of the effort that is going to be needed (Bowker, 2000; Watson-Verran & Turnbull, 1995).

To integrate knowledge across different fields, experts must spend time and effort sharing knowledge and creating common understandings. Active discussion and engagement are essential strategies for achieving this (Clark, 1996; Cook & Brown, 1999; Engestrom, 1999; Klein, 1990), yet one of the major challenges faced is providing the operative infrastructure that allows experts sufficient time to bring their expertise to bear on a problem. Studies summarized by Bransford et al. (1999) show that experts' apparent fluent retrieval of the appropriate knowledge to bring to bear on a problem does not mean it takes them less time than novices to solve problems. Rather, we are faced with the "mythical man-month" problem (Brooks, 1975)—communication time increases exponentially with the addition of more experts. Duncker-Gassen (1998) demonstrates how this occurs in practice: The experts need to develop a common language in which just enough and not too much of their internal expertise is communicated at team meetings. It is not surprising, then, that at a recent NSF workshop for principal investigators on interdisciplinary projects, a common lament was that everything took more time than expected (NSF KDI workshop, April, 2002; for a report see Cummings & Kielser, 2004).

In addition, changes in disciplinary practices also continue to meet barriers created by institutional practices. In interviews with research team members, and in comments in many academic venues, university promotion and tenure (P&T) committees are repeatedly named as arbiters of academic practice, and thus weigh heavily on the minds of interdisciplinary researchers, particularly team leaders and mentors who have the interests of their doctoral students and untenured faculty in mind. P&T committees have the power to make or break an academic career, and thus what they accept in terms of academic practice significantly shapes and constrains research activity. Several issues—including standards about co-authorship, publication venues and legitimate research methods or subject areas—may limit on what and with whom group members will work, as well as where and how work will be expressed and distributed. Recent attention has been given to the acceptability of electronic publications. Kling, McKim and King (2003) report that at a workshop on electronic journals at an information systems conference, many participants, and in particular more junior scholars, reported they were afraid to publish in new electronic scholarly communication forums because of a concern that "their Promotion and Tenure (P&T) committees would not count these publications or would disapprove of them in promotion decisions" (p. 67). This problem is even greater for humanities scholars (and some social scientists) who engage in work with new technologies. Their challenge can be even greater because collaborative activity is not yet regarded as acceptable to many promotion committees. Those combining new computer developments with collaborative and interdisciplinary activity then face a further hurdle as they try to find outlets for publication and recognition of their work (Greenblatt, 2002).

Similarly entrenched notions of how grant funding is given, for what, and for how long, may hamper the efforts of those working on and examining infrastructure development. Kling, et al. (2003) describe how the use of FlyBase, a database about the *Drosophila* (fruit fly) genome, ran up against this problem.

FlyBase is not funded as an NIH [National Institute of Health] line item— rather, the coalition must apply for new grants every 3-5 years. NIH research grants are generally not given for maintenance and operations, but rather for new developments. Therefore, the FlyBase coalition must continually propose the development of new features to receive funding for continued operation. (Kling et al., 2003, p. 62)

A social science group we studied faced the same kind of problem at the end of its first grant funding. Having developed an environment for evaluating, maintaining and disseminating educational materials, they reached the end of their grant and were left searching for ongoing support for maintenance and further development.

Thus, we see how conceptions and preconceptions about what constitutes disciplinary practice can limit interdisciplinary work, and how existing social, technical and institutional frameworks may conflict with the emergent nature of interdisciplinary efforts, as well as emergent, continuing needs.

Practices within Teams

Practices within teams are also often overlooked in measuring the complexity of interdisciplinary collaboration. While all teams face difficulties in pooling and sharing knowledge and making the most of member contributions, interdisciplinary teams face particular challenges around their lack of redundancy in disciplinary coverage and their lack of a shared base of domain and procedural knowledge. Interdisciplinary, collaborative teams have *knowledge interdependencies* that may be overlooked or oversimplified if a mono-disciplinary model of work is assumed. These interdependencies manifest themselves in several ways. As a group, interdisciplinary teams depend on members to represent their disciplines and to bring that knowledge to the team's work. Consequently, members representing that discipline must take on the responsibility to be present and contribute continually to the overall endeavor. As a corollary, those from other disciplines must be open to hearing and incorporating such knowledge. To integrate and synthesize knowledge, the group must be ready to engage together in contributing knowledge and learning from others.

Learning as a group may be mutual as members gain new knowledge together—what might be described as *co-learning* (e.g., through joint discussion of papers read in common, bringing in experts to consult with the team and attending conferences together). Individuals may also learn on behalf of the group, e.g., by expanding their knowledge of another's area, or even engaging in *altruistic learning,* as individuals learn for the sake of the group. For example, individuals might take classes or visit other laboratories to bring expertise in a software program or research technique into the group. Strategies to promote such learning may include reading about the content or methods of another's discipline, doing background research to gain knowledge in new common areas or learning with the intent of training others in the group (for more on learning in groups, see Argote, Gruenfeld & Naquin, 2001).

Learning also entails getting to know other members of the team, and what they know. Members form *transactive memory* (i.e., memory about "who knows what"; Moreland, 1999; Wegner, 1987) about each other's knowledge, expertise, methodologies and approaches, working styles, available time and commitment. They acquire knowledge about each other's disciplines, come to understand what work may forward an individual's career and how these constraints play out in an individual's needs and urgencies (e.g., to finish a dissertation in a year; to publish for tenure packages). Team members also learn who does what tasks and who has what responsibilities (Brandon & Hollingshead, 1999). Once "directories" of who knows what are in place (whether in the heads of team members, or held externally in knowledge management systems), team members can reduce their own information load and responsibility by forwarding information and/or questions to appropriate experts (Monge & Contractor, 2003; Palazzolo, 2003; Wegner, Giuliano & Hertel, 1985). Such learning also builds common ground as group members create a collective or *community memory* that defines them (Orr, 1996; Chayko, 2002; see also the section on "Seeing and Crossing Boundaries").

However, these learning processes may also involve shedding existing biases and predispositions regarding what others do in their domain. We may begin with projected notions of what others do and think as computer scientists, biologists, sociologists, psychologists, etc. Biases include stereotypes about personalities, data collection practices, attitudes toward study participants, analysis skills and technology skills and interests. Yet, these stereotypes rarely hold, and someone representing one discipline may have spent a long time working in another and thus be expert in several fields and approaches. Indeed, our own group, which we present as a whole as composed of social scientists, includes PIs with academic degrees in computer science, education, history, information science, management, philosophy and psychology; home departments of communication, education, library and information science, and management; work experience both in and outside academia; and research endeavors from sociology of science to the development of educational technology.

Preconceptions about people may also apply to projected notions of their willingness to learn. Novices may be expected to learn widely, yet degree requirements for graduate students may focus them more narrowly on their own discipline. Senior experts, by contrast, may be thought of as narrowly defined, yet by taking on the interdisciplinary endeavor may be learning more widely. Sharon Traweek's groundbreaking work on the high-energy physics community is key here (Traweek, 1992): the senior physicist tends to believe more in (and has more to gain from) active collaboration than the junior researcher who has yet to stake his own claim to a piece of the field (cf. Star & Ruhleder, 1996). The expert who acts as an "accomplished novice" may come from any field, yet junior participants such as students may never before have experienced cooperative and collaborative practices with senior experts, and thus have neither expectations nor practical experience with such practices. Because of their invisible nature, we may also forget to include training in interdisciplinarity and collaborative methods when we introduce new people into a team.

Preconceptions also include assumptions of who will work with whom, because of their age, discipline, home institution, etc. Often graduate students are expected to talk and work with each other, yet this can be an added burden when they are only beginning to gain expert grounding in their own discipline. To talk across disciplines requires a solid understanding of typical practices such as collecting and analyzing data and making appropriate claims, as well as the ability to think abstractly and metacognitively about how combining widely disparate disciplinary efforts may challenge and redefine one's home practices. Many PIs at the NSF KDI workshop commented on how their students had not yet acquired sufficient disciplinary skills when they were plunged into interdisciplinary work. Thus, PIs need to judge when their graduate students are as ready for interdisciplinary work as they are for disciplinary work.

When assumptions are made about others' work, there may be preconceptions about *how collaboration will proceed*: for example, what kind of information or data will be made available to all members of the group, when this will occur, what pre-work enables that sharing and whether data can be shared across scales, methods, human subjects permissions, etc. We also have assumptions about *how work will progress*, for example, from setting goals to designing studies, collecting data and producing papers. If team members tacitly anticipate a certain set sequence of research design, data collection, analysis and publication, they may ignore the more complex logistics necessary for collaborative work, such as the pooling of data and resources (equipment, bibliographic citations, etc.), as well as the appropriate synchronization of different tasks. Even if team members share expectations about a project's goals and progression, how do these match the expectations of the external organizations that house and/or fund them? How can they convey new, evolving, interdisciplinary practices and outcomes to funding agencies, participant organizations, advisory boards, universities and departments?

In our own work we have found several instances where expectations were out of synch with what was possible, or what others thought. In data collection, we found that research participants prejudged their role in our research, e.g., declining interviews because they "didn't know anything about interdisciplinary activity," when we wanted to hear about what they did. Similarly, we found that participants prejudged our aims as "evaluative" and thus feared an impact on their future grants. We find we are not alone in being judged this way: Suchman, Blomberg, Orr and Trigg (1999) report how they took "great pains" to explain the point of their observations and recordings of work processes in an organizational context, where "it is difficult for people who have grown up in the shadow of scientific management to imagine what interest researchers could have in their work other than to evaluate it in terms of workers' competency and efficiency" (p. 398).

Likewise, in managing our own collaboration, the co-PIs spent over a year meeting regularly in preparing the grant application and developing common consensus on research approaches. Yet, once we got the grant, graduate students were brought on and expected to "just know" about the collaborative approach the PIs had spent so much time showing each other. We also failed to recognize assumptions about different methods we were using, including how well results from different methods could be timed to feed into other parts of the overall work (e.g., ethnographic results into social network studies). As we tried to put work together, we found our different approaches came with different and sometimes conflicting needs. Members have had to negotiate expectations for defined versus open-ended study populations, hypotheses over more open-ended research questions and results verified by statistical tests versus philosophical reasoning (e.g., statistical testing of the impact of membership on collaborative grants to numbers of publications, versus open interviews coded based on principles of grounded theory versus philosophical analyses of the nature of collaboration). Each of these has represented challenges to our own work that are also likely to be faced by others.

Seeing and Crossing Boundaries

A further challenge to interdisciplinary endeavors is the time and commitment members give to a particular project. Experts have their operations in other fields, and spend only limited time in the interdisciplinary endeavor. They have different notions of their involvement than do outside observers.

The challenge of boundaries became apparent to us when we tried to define who belonged to the groups we wanted to study. Despite the existence of web sites listing contacts, grant funding agency documents with PI names, and lists of employees in organizations, this was not an easy task. We found ourselves asking: What constitutes a person, a collective, an institution? There are answers which everyone knows intuitively, but we found the entities ever more ill-defined as we tracked them acting in our teams. Within the distributed projects we studied, every collective designation proved problematic. Who were the members of the research team? It certainly was not the list of people to whom the grants were awarded—members dropped out and came on board over time—so that the organizations we were studying (NCSA and the home institutions of the grantees) often had different accounts of membership. PIs receiving only small percentages of funding from a particular grant would remind us of where this fit in the larger arena of their research. We had to ask whether 5% funding led a PI to consider him or herself committed to the project (and if so to what extent), and whether this level of involvement meant they counted as a "person" or "unit" in terms of group membership. Is the definition of the collective "group" based on those with some minimal time or funding involvement, or does it consist of the names on a web site? Is an organization involved if only one member of their staff, now departed, used to work on the project?

A similar problem came when considering how team members represented a particular discipline. Does an individual represent computer science (CS) only if they have a degree in CS? What if they are working on a CS degree but don't yet have one? What if they had worked in computing but had no CS degree? What if their degree is in CS, but they work in a different department—which discipline would we say they represented? What if they published in CS journals, or about CS in non-CS journals—does that make their discipline CS? The answer probably lies somewhere between "all of the above" and "none of the above," that is, somewhere in the interdiscipline.

Intra-Team Boundaries

How individuals set boundaries for themselves, partitioning and allocating time to projects can significantly impact scheduling and coordination. Scientists involved in several grants may apportion time and commitment based on which team demands the most time, which team is just in the beginning stages of work (or other critical stages), which team members are most local and therefore most visible for reminders about work or which working arrangement or topic is more prestigious or popular. Within the project, teams negotiate the boundaries of work, including roles, responsibilities, project scope, and intended outcomes. They set *temporal* boundaries, such as when the project will end (or, more instrumentally, when the funding will run out), and *trajectories* for the work, essentially mapping out future boundaries against which progress will be measured.

Interdisciplinary teams learn to partition and allocate time, money, and resources, as well as access to equipment, software, data, reports, and people. For example, when studying the same group from multiple perspectives, we needed to manage our own access to respondents. We considered how to space requests over time, and limit access to a team to only one set of researchers at a time, much as others might book time on equipment or for the use of computing resources. Researchers thus may be seen to mark *territory,* for example, deciding who studies who and when, determining hand-over points and coordination schedules. As researchers, we became aware of distinctions between our disciplines based on differences in methods and approaches, and attitudes to the relationship between the researcher and the participant (e.g., the invisible observer or arms-length researcher versus the involved, participant observer).

Individual Attitudes to Boundaries

Group members also consider the boundaries to their own identification: What discipline do they feel they speak for when working on an interdisciplinary team? How do they explain their work to their home discipline (however defined)? How do they describe their work when it is presented for tenure and promotion? Many individuals on the teams we have studied have been acutely aware of this problem of crossing boundaries.

Team members may also differ in their need to define boundaries for their work, and whether their own perspective sets boundaries (or eschews them). For example, some researchers may not be comfortable unless the boundaries of the study population are set—and hence seek definition of such boundaries by creating operational definitions of inclusion and exclusion (e.g., being on the payroll, listed on a web site, belonging to a department; or bounded by geography, politics or topology). Others may accept a more emergent, even amorphous boundary, and hence see emergent definitions of involvement (e.g., by using "snowball" techniques that begin with a set of people and ask them who to talk to next; or by letting boundaries emerge from observation). Comfort with this definition, and even the need for a definition, affects how team members view the legitimacy of the work.

Goals as Boundaries

Goals act as boundaries, orienting, but also constraining, activities. Our research suggests that interdisciplinary activity particularly needs to accommodate change and adjust goals. Indeed, we argue that the goals of interdisciplinary groups *need* to evolve so they do not become barriers instead of boundaries.

In defining boundaries, creating schedules, marking territory and establishing trajectories early in an interdisciplinary project, groups can close discussion too early; discussion based on a non-confrontational, "illusion of friendship" (Klein, 1990) may lead to quick agreement, missing the opportunity to learn from each other, jointly "formulating and debating a problem" (Engestrom, 1999, p. 380), setting a joint agenda and generating new knowledge (Cook & Brown, 1999). Not only is knowledge about the interdisciplinary endeavor at an early stage, and should be kept open for the generation of new knowledge (Cook & Brown, 1999; Haythornthwaite, Lunsford, Kazmer, Robins, & Nazarova, 2003), so should the tasks and outcomes be open for continuous negotiation. While all groups have work to achieve—and as such need to come together around a common goal, and to create and determine schedules and processes—interdisciplinary groups, especially those forging new interdisciplinary relations, face the particular challenge that outcomes of their work may not be apparent when they begin working together. Too many approaches to interdisciplinary work start with the premise of a predetermined, fixed end to be achieved. The organization of work that ensues can inhibit collaborative interchange. For example, in a collaboratory for structural biology studied by one of our members, well-defined roles and schedules existed for gathering information on what the user community required (first stage of project), coding that requirement into collaboratory tools (second stage) and then evaluating the overall result (final stage). While these well-defined roles produced a collaboration of sorts, the coordinated activity reduced two-way collaboration, for example, participants entering late in the schedule were not working together with those whose roles entailed earlier work. What results is not collaboration, but work characterized by coordination, as pieces are handed over as completed, thus missing the potential benefits of fully collaborative interaction.

Interdisciplinary work may be particularly difficult precisely because *goals emerge from the interaction, not prior to it*. We believe that interdisciplinary, collaborative work requires *continuous negotiation and co-construction of outcomes*, which further entails continuously engaging individuals around the definition of an emergent and mutually agreeable result from the work (see also Haythornthwaite et al., 2003).

Defining parameters early may also lead to misconceptions about what other team members know or will do. Our knowledge about group members is at a formative stage when we begin to work together, as is our knowledge of what tasks others are responsible for and what they know (Brandon & Hollingshead, 1999; Wegner, 1987). As an outsider looking in on knowledge processes, we may fail to acknowledge the side-tasks and invisible work (Star & Strauss, 1999) that are so important to group functioning. Just as many early management literatures focused on "process losses" and failed to give weight to the social relations that keep members committed to the group (McGrath, 1984), so, too, a focus on interdisciplinary knowledge work alone fails to recognize other corollary knowledge exchanges. Results from the scientific teams we studied (Haythornthwaite & Steinley, 2002; Haythornthwaite, forthcoming) suggests that a number of apparently extraneous kinds of knowledge represent important exchanges among team members, including advice on how to work with funding agencies, knowledge of who outside the team can provide personal references or knowledge about future jobs for graduate students. From our own experiences, we know the importance of gaining knowledge about how to communicate via video and audio-conferencing. Seemingly inconsequential issues, such as learning how to accommodate temporal delays in videoconferencing and how to include invisible phone participants, reflected very important gains in our ability to accomplish work and communication at a distance. (These issues are addressed further in the "Working With and Through Technologies" following section.)

Managing External Relations

Managing external relations may be considered a corollary to the discussion of boundaries given above, but in this section, attention is given to what is outside rather than inside the boundary. It is important for teams to recognize and manage both individual and group relationships with these external entities. As Ancona and Caldwell (1990) note:

To be successful, new product teams must obtain information, resources, and support from others, both inside and outside of the organization, use that information to create a viable product, and finally transfer the technology and enthusiasm for the product to those who will bring it to market. . . . This makes the new product team highly dependent upon others, and suggests that an important way of understanding the performance of these teams lies in examining how they manage relationships with other groups. (Ancona & Caldwell, 1990, p. 120)

New interdisciplinary teams, like new product teams, have ideas, projects and careers to sell to granting agencies, home departments and universities.

Teams, particularly grant-funded teams operating in departments in research institutions, manage a number of layers of external relations, some of which have been mentioned above: home departments and colleges, university (or other institution) administration, review boards (e.g., Institutional Review Boards (IRBs) that sanction the ethics of the research), and granting agencies. Teams and team members often need to manage an image of themselves that is then often used again to promote the external agencies. Thus, teams may liaise with the public through brochures, pamphlets, web sites and interviews. Teams are also seen as representatives of such agencies, and may be required to identify their association under some circumstances, as well as, paradoxically, disclaiming the agencies' involvement with the results that are presented (e.g., through the addition of lines such as "The findings presented in this report do not necessarily reflect the views of the <fill in the blank> agency").

Team members also act as representatives of interdisciplinary work in their home disciplines. Thus, legitimate responsibilities for such work includes taking information, knowledge, practices and methods *back* to home disciplines where they may be used by others (Kanfer et al., 2000). This may pave the way for the next generation of interdisciplinarians in a form of legitimate peripheral participation (Lave & Wenger, 1991), as well as legitimizing the interdisciplinary endeavor, and possibly fostering new practices in the home disciplines.

Working with and through Technologies

As contemporary groups work and forge new synergies, they typically do this with attention to the adoption, integration and creation of new computing technologies. Both funding initiatives and peer pressure in many settings encourage teams to adopt new technologies to support distributed knowledge processes and creative work. We separate out our discussion of them here because they add another, often unrecognized, cognitive and logistic load on contemporary teams, as new technologies require the development of new practices and the reinvention of old ones.

Bridging Practices with Technology

To bridge differences in disciplinary practice, teams often define social and technical infrastructures to support their joint work. Many contemporary teams include as part of their mandate the creation of new technologies relating to their research endeavor, such as data repositories, software, analysis techniques, digital libraries and collaboratories for shared data collection and analysis. This work entails both defining protocols, data units, database structures, computer interfaces, storage and retrieval techniques, etc., as well as encouraging joint use of these technologies. In the same manner as for learning, members may need to engage in altruistic use of technologies, seeding shared databases to establish a critical mass of data or communication that makes the effort worthwhile for others (Markus, 1990; Connolly & Thorn, 1990).

However, it is at the definitional interfaces that interdisciplinary teams in particular meet many challenges. Disciplinary standards on data units, naming conventions and analysis techniques must be generalized to cover many disciplines. Those who study processes that unfold over millennia try to match up with data from those who study yearly phenomena. Those who work with quantitative data and approaches try to match up with qualitative data and approaches. Moreover, long-held definitions about what a concept means or a thing "is" have established infrastructures that affect what kinds of questions are asked by certain disciplines (see Bowker & Star, 1999). Likewise, evolving definitions for new entities, such as "collaboratories," affect what counts as the object of study by different disciplines (see Lunsford & Bruce, 2003).

To illustrate the complexity inherent in the meaning of a thing, consider bringing experts together to work on a simple thing like soil (Gray, 1980; Bowker, 2000). Different sets of researchers will have very different intuitive definitions of what soil is. An agriculturalist views soil as something crops grow in; ecologists see soil as including hard rock (to which lichens cling), and as part of the earth's surface subject to weathering influences. Such disciplinary views suggest specific ways of looking at soil, and each suggests different kinds of questions to ask. What constitutes the "universe" of soil differs by classification scheme. For example, soil classification schemes used for North America and Europe include national variations found in those countries. However, since these countries lack a tropical climate, soils of the tropics do not get detailed coverage in these classifications. Thus, geographic and interdisciplinary differences can predispose researchers to a particular view of a supposedly common object. In interdisciplinary work, definitions need to be reopened and compared across fields, adding to the overhead of such work, the time it takes to complete projects, and the uncertainty regarding what the project is about.

Importantly, while systems development is rapid, change in practice is slower and more complex. Practice is bound up with traditions within fields, entrained with funding cycles and the need to show results. You cannot, as the Flora of North America (http://www.fna.org/FNA/) project discovered, easily convince botanists to move from traditional glossy plates and printed text to electronic publication of only partially-vetted results. Change in culture is slow even though this kind of advance is doubly useful as it speeds up work within the botanical community, and opens the borders of that community for researchers in cognate disciplines to tap into ongoing work more readily. Similarly, the Long Term Ecological Research (LTER) network (http://lternet.edu/) has found difficulty trying to convince researchers to produce truly interoperable databases: Every-one agrees it would be valuable; however, getting scientists to think outside their current project (and funding cycle) is extremely difficult.

Comparable difficulties are found in spanning disciplines. For example, the new field of systems biology, which melds computer science, engineering and biology, entails a change from analyzing "one gene or protein at a time" to examining "a living thing as a whole":

Doing systems biology requires a huge change in the research culture. . .In traditional molecular biology, each scientist works on his own gene, but the systems approach requires determining the effect of every gene on every other. You have to give up this 'my gene, your gene' stuff. (Eric Davidson, quoted in Begley, 2003, B1)

A general rule of development to date has been that domain scientists have been left to deal with the cultural change on their own, with input from frustrated systems developers who wonder why their programs are not being used (Star & Ruhleder, 1996).

Similarly, different fields have different ideas about when data should be made public, for example, when collected, after publication of results or not at all. There are different standards on publication venues (online or offline; proceedings, journal article and/or book), and authorship practices (single or multiple; including or excluding research assistants). Moreover, such approaches are often so intricately tied to what it means to be a biologist, computer scientist, or psychologist that it is extremely difficult to accept another's methods as legitimate practice, *and* to have those practices accepted across fields and departments. Interdisciplinary groups must make integrative efforts to learn and communicate not only in combining what is known in a field, but also in bringing together their various approaches and forms of data collection and handling, and applying technologies that suit both individual needs and joint work. Crossing such divides, reopening definitions and even recognizing that differences exist, add to the challenges of interdisciplinary work (see also Haythornthwaite, 2004).

Bridging Distance with Technology

As developments in information and communication technologies make it possible and expected that we can work at a distance (anytime, anywhere), we find that we are paradoxically more fragmented in our communication channels at the same time as endeavoring to be more integrative in our practices. Not only must groups create new technical infrastructures to bring their different work together, they also learn "side" activities associated with accommodating distributed work practices. They must coordinate work schedules across time and geography, establish norms for interpersonal communication via new media and keep up with the ramifications of the everyday presence of information and communication technologies (such as on and offline publishing standards and the need for representing the team's research on the internet).

Finding opportunities to communicate are more difficult for teams that cross departments, institutions and geography since their activities do not have the same entrainment (McGrath, 1990) to local environments (e.g., the local coffee spot), events (e.g., faculty meetings, pub nights), schedules (e.g., semester cycles, holiday breaks) and time zones (e.g., common work hours). Although it may seem a trivial point, such lack of common time can seriously hamper scheduling joint activities. For example, in our own group, with members from the east to the west coasts of the U.S., we have had to find meeting times that do not have our west coast participants up at 6 a.m., and yet do not occur during anyone's core hours of the day when their own local meetings and classes take place. Further, operating on different semester systems means our break times do not regularly overlap, and so face-to-face meetings are shoehorned into what overlap does exist, often meaning long delays between such meetings. Thus, we have found that we spend time learning about the schedules and daily operating routines of institutions other than our own. (See also Livia, 1999, who describes how time zone differences between herself and her research partners and participants affected doing research via the internet.)

Distributed groups may find that part of their learning to work together involves experimenting with computer-mediated communication (CMC) and work tools for distributed meeting support, as our group has. Asynchronous tools, such as e-mail, modify the way individuals communicate information and coordinate their actions—paradoxically both enhancing and constraining communication (for reviews, see Kiesler & Sproull, 1992; Wellman, Salaff, Dimitrova, Garton, Gulia & Haythornthwaite, 1996; Haythornthwaite, Wellman & Garton, 1998; Herring, 2002). For example, the reduced social cues in an e-mail message may constrain coordination efforts by failing to engender commitment to a course of action. Overload in e-mail may lead to important messages being lost in a sea of other communications; and social gatherings may be curtailed because of a perception that e-mail (or other forms of CMC) is "as good as being there." At the same time, e-mail communications may provide an ideal way to keep everyone in the loop on activities, and provide an easy, central means of coordinating action and recording activities.

Our own experiences with communicating via a combination of conferencing media highlight some ways in which technologies modify group dynamics, and how we had to learn to adjust to them. A common setup for us for most of the project has been to have a set of people face-to-face in a conference room, a video display bringing in participants from one to four sites across the U.S. and audio-conferencing for remote participants without video capabilities. We have found that distributed group members who join meetings by videoconference, while visible and obviously present, are disadvantaged in conversations because the one to three second transmission delays are just long enough for non-delayed speakers to jump in first. Phone participants, on the other hand, are physically invisible and in danger of being forgotten as present at the meeting. The presence of others in the local meeting room (seen face-to-face) and/or on the larger-than-life video display, overshadows these invisible listeners. Over time, we have learned to pause the requisite one to three seconds for responses from video-attendees, to poll those "in" the phone for contributions, to voice a "yes" rather than nodding and to attend to more tacit signs of presence, such as an additional breathing over the phone, or sounds that indicate a new person has dialed into the conference system.

As groups look to technology to support their work processes, we add the caveat that technology (such as videoconferencing) does not provide a "solution" to the "problem" of collaboration. Instead, we see technology both emerging from and modifying practice, leading on to the next generation of problems and solutions. This *pragmatic technologies* view (Hickman, 1992) allows us to see technologies as both antecedent and consequence of distribution, interdisciplinarity and collaboration, and groups as able to both adopt and adapt technologies to their needs (Bruce & Hogan, 1998). Similarly the *adaptive structuration* view (Descantis & Poole, 1994) lets us see social processes emanating from group uses of a technology rather than some standard across-the-board definition of use. Like interdisciplinary outcomes, technology use by groups and for groups needs to be seen as developing from use, and to recognize that such changes and emergent use are important aspects of interdisciplinary, collaborative practices.

Conclusion

We have brought together here some challenges characterizing distributed, interdisciplinary, collaborative practice, encompassing learning and communicating, preconceptions and expectations, boundaries, external relations and social and technical infrastructures. Although many groups, interdisciplinary or not, collaborative or not, will face the challenges discussed here, we find that another challenge for distributed, collaborative groups is one of *quantity*: At every turn there is another aspect of group operation, interaction and endeavor that must again be worked out from scratch. Education in a single field does not prepare us for the many factors that drive and sustain successful interdisciplinary, collaborative work.

With this long list of challenges, we are concerned lest the reader feel that we have established a wall of obstacles to interdisciplinary progress. To the contrary. There are good research and structural reasons why agencies should fund and support integrative efforts. Chief among them, perhaps, is that the issues we face in managing the planet (climate warming, biodiversity preservation) and in managing complex social and political decision making (the tribulations of a globalizing world) necessarily call forth interdisciplinary responses. However, we believe the success of such endeavors will depend on recognizing the characteristics and challenges to such work, and acting to support this new kind of work. Moreover, the challenges must be recognized as both technical and social: Doing technically integrative, interdisciplinary work represents a major change in the way people work together, what problems they address and how and what products they complete as a result. Yet it remains an opportunity to be pursued.

Acknowledgment

The authors have been supported in their examination of distributed collaboration by a National Science Foundation, Knowledge and Distributed Intelligence grant (award 9980182). This chapter has benefited greatly from participation by Haythornthwaite and Bowker in a working group on scientific collaboration supported by the National Center for Ecological Analysis and Synthesis, Santa Barbara, CA. The findings do not necessarily reflect the views of these agencies. We also thank the members of the DKRC project for their input: Michelle M. Kazmer, Joyce Brown, Michael DeVaughn, Harald M. Fischer, Nicholas Burbules, Alaina Kanfer, Sarai Lastra, Tim McDonough, Joseph F. Porac, Katie Vann and James B. Wade.

References

Ancona, D. G., & Caldwell, D. (1990). Beyond boundary spanning: Managing external dependence in product development teams. *The Journal of High Technology Management Research, 1*(2), 119-135.

Argote, L., Gruenfeld, D., & Naquin, C. (2001). Group learning in organizations. In M. E. Turner, *Groups at work: Theory and research* (pp. 369-411). Mahwah, NJ: Lawrence Erlbaum.

Begley, S. (2003). Biologists hail dawn of a new approach: Don't shoot the radio. *The Wall Street Journal.* Friday, February 21, B1.

Bowker, G. C. (2000). Biodiversity dataversity. *Social Studies of Science, 30*(5), 643-684.

Bowker, G. C., & Star, S. L. (1999). *Sorting things out: Classification and its consequences.* Cambridge, MA: MIT Press.

Brandon, D., & Hollingshead, A. (1999). Collaborative learning and computer-supported groups. *Communication Education, 48*(2), 109-126.

Bransford, J. D., Brown, A. L., & Cocking, R. R. (1999). (Eds.), *How people learn: Brain, mind, experience, and school.* Committee on Developments in the Science of Learning Commission on Behavioral and Social Sciences and Education, National Research Council, Washington, DC: National Academy Press. Retrieved July 21, 2005, from http://www.nap.edu/html/howpeople1/

Brooks, F. (1975). *The mythical man-month: Essays on software engineering.* Reading, MA: Addison-Wesley.

Bruce, B. C., & Hogan, M. C. (1998). The disappearance of technology: Toward an ecological model of literacy. In D. Reinking, M. McKenna, L. Labbo, & R. Kieffer (Eds.), *Handbook of literacy and technology: Transformations in a post-typographic world* (pp. 269-281). Hillsdale, NJ: Erlbaum.

Chayko, M. (2002). *Connecting: How we form social bonds and communities in the internet age.* Albany: State University of New York.

Clark, H. (1996). *Using language.* Cambridge, UK: Cambridge University Press.

Connolly, T., & Thorn, B. K. (1990). Discretionary data bases: Theory, data and implications. In J. Fulk & C. W. Steinfield,(Eds.),*Organizations and communication technology* (pp. 219-234). Newbury Park, CA: Sage.

Cook, S. D. N, & Brown, J. S. (1999). Bridging epistemologies: The generative dance between organizational knowledge and organizational knowing. *Organization Science, 10*(4), 381-400.

Cummings, J., & Kielser, S. (2004). *KDI initiative: Multidisciplinary scientific collaboration.* Retrieved July 21, 2005, from http://netvis.mit.edu/papers/NSF_KDI_report.pdf

DeSanctis, G., & Poole, M. S. (1994). Capturing the complexity in advanced technology use: Adaptive structuration theory. *Organization Science, 5*(2), 121-47.

Duncker-Gassen, E. (1998). *Multidisciplinary research at the University of Twente.* Enschede: Twente University Press.

Engeström, Y. (1999). Innovative learning in work teams: Analyzing cycles of knowledge creation in practice. In Y. Engeström, R. Miettinen, & R. Punamäki (Eds.), *Perspectives on activity theory* (pp. 377-404). Cambridge, UK: Cambridge University Press.

Fienberg, S. E., Martin, M. E., & Straf, M. L. (1985). *Sharing research data.* Washington, DC: National Academy Press.

Gray, B. (1980). Popper and the 7th approximation: The problem of taxonomy. *Dialectica, 34*(2), 129-153.

Greenblatt, S. (2002). A letter to MLA members. *Chronicle of Higher Education,* Tuesday, July 2, 2002. Retrieved July 21, 2005, from http://chronicle.com/jobs/2002/07/2002070202c.htm

Haythornthwaite, C. (2004). *Communicating knowledge: Articulating divides in distributed knowledge practice.* New Orleans, LA: International Communication Association.

Haythornthwaite, C. (forthcoming). Learning and knowledge exchanges in interdisciplinary teams. *Journal of the American Society for Information Science and Technology.*

Haythornthwaite, C., Lunsford, K. J., Kazmer, M. M., Robins, J., & Nazarova, M. (2003). The generative dance in pursuit of generative knowledge. In *Proceedings of the 36th Hawaii International Conference on System Sciences.* Los Alamitos, CA: IEEE Computer Society.

Haythornthwaite, C., & Steinley, D. (2002). *Interdisciplinary knowledge exchange.* Paper presented at the International Sunbelt Social Network Conference, New Orleans, LA.

Haythornthwaite, C., Wellman, B., & Garton, L. (1998). Work and community via computer-mediated communication. In J. Gackenbach (Ed.), *Psychology and the internet* (pp. 199-226). San Diego, CA: Academic Press.

Herring, S. C. (2002). Computer mediated communication on the internet. *ARIST, 36,* 109-168.

Hickman, L. A. (1992). *John Dewey's pragmatic technology.* Bloomington: Indiana University Press.

Kanfer, A., Haythornthwaite, C., Bowker, G. C., Bruce, B. C., Burbules, N., Porac, J. F., & Wade, J. B. (2000). Modeling distributed knowledge processes in next generation multidisciplinary alliances. *Information Systems Frontier, 2*(3/4), 317-331.

Kiesler, S., & Sproull, L. (1992). Group decision making and communication technology. *Organization Behavior and Human Decision Processes, 52,* 96-123.

Klein, J. (1990). *Interdisciplinarity: history, theory, and practice.* Detroit, MI: Wayne State University.

Klein, J. T. (1996). *Crossing boundaries: knowledge, disciplinarities, and interdisciplinarities.* University Press of Virginia.

Klein, J. T., Grossenbacher-Mansuy, W., Haberli, R., Bill, A., Scholz, R. W., & Welti, M. (2001). *Transdisciplinarity: Joint problem solving among science, technology, and society.* Switzerland: Birkhauser.

Kling, R., McKim, G., & King, A. (2003). A bit more to it: Scholarly communication forums as socio-technical interaction networks. *Journal of the American Society for Information Science, 54*(11), 47-67.

Knorr-Cetina, K. (1999). *Epistemic cultures: How the sciences make knowledge.* Cambridge, MA: Harvard University Press.

Lave, J., & Wenger, E. (1991). *Situated learning: Legitimate peripheral participation.* Cambridge: Cambridge University Press.

Livia, A. (1999). Doing sociolinguistic research on the French Minitel. *American Behavioral Scientist, 43*(3), 422-435.

Lunsford, K. J., & Bruce, B. C. (2003). Collaboratories: Working together on the web. In B. C. Bruce (Ed.), *Literacy in the information age: Inquiries into meaning making with new technologies* (pp. 290-299). Newark, DE: International Reading Association.

Markus, M. L. (1990). Toward a "critical mass" theory of interactive media. In J. Fulk & C. W. Steinfield (Eds.), *Organizations and communication technology* (pp. 194-218). Newbury Park, CA: Sage.

McGrath, J. E. (1984). *Groups, interaction and performance.* Englewood Cliffs, NJ: Prentice-Hall.

McGrath, J. E. (1990). Time matters in groups. In J. Galegher, R. E. Kraut, & C. Egido (Eds.), *Intellectual teamwork: Social and technological foundations of cooperative work* (pp. 23-61). Hillsdale: Lawrence Erlbaum Associates

Monge, P. R. & Contractor, N. S. (2003). *Theories of communication networks.* Oxford, UK: Oxford University Press.

Moreland, R. (1999). Transactive memory: Learning who knows what in work groups and organizations. In L. Thompson, J. Levine, & D. Messick (Eds.), *Shared cognition in organizations* (pp. 3-31). Lawrence Erlbaum Associates.

Orr, J. (1996). *Talking about machines: An ethnography of a modern job.* Ithaca, NY: Cornell University Press.

Palazzolo, E. (2003). *Structures of communication to retrieve information in organizational work teams: A transactive memory perspective.* Unpublished doctoral dissertation, University of Illinois at Urbana-Champaign, Urbana, Illinois.

Porac, J. F., Wade, J. B., Fischer, H. M., Brown, J., Kanfer, A., & Bowker, G. C. (2004). Human capital heterogeneity, collaborative relationships, and publication patterns in a multidisciplinary scientific alliance: A comparative case study of two scientific teams. *Research Policy, 33*(4), 661-678.

Scardamalia, M., & Bereiter, C. (1996). Computer support for knowledge-building communities. In T. Koschmann (Ed.), *CSCL: Theory and practice of an emerging paradigm* (pp. 249-268). Mahwah, NJ: Lawrence Erlbaum.

Star, S. L., & Ruhleder, K. (1996). Steps toward an ecology of infrastructure: Design and access for large information spaces. *Information Systems Research, 7*(1), 111-134.

Star, S. L., & Strauss, A. (1999). Layers of silence, arenas of voice: The ecology of visible and invisible work. *CSCW, 8*(1-2), 9-30.

Suchman, L., Blomberg , J., Orr, J., & Trigg, R. (1999). Reconstructing technologies as social practice. *American Behavioral Scientist, 43*(3), 392-408.

Traweek, S. (1992). *Beamtimes and lifetimes: The world of high energy physicists*. Cambridge, MA: Harvard University Press.

Watson-Verran, H., & Turnbull, D. (1995). Science and other indigenous knowledge systems. In S. Jasanoff, G. Markle, J. Petersen, & T. Pinch (Eds.), *Handbook of science and technology studies* (xv, 820). Thousand Oaks, CA: Sage Publications.

Wegner, D. (1987). Transactive memory: A contemporary analysis of the group mind. In B. Mullen & G. Goethals (Eds.), *Theories of group behavior* (pp. 185-208). New York: Springer-Verlag.

Wegner, D., Giuliano, T., & Hertel, P. (1985). Cognitive interdependence in close relationships. In W. J. Ickes (Ed.), *Compatible and incompatible relationships* (pp. 253-276). New York: Springer-Verlag.

Wellman, B., Salaff, J., Dimitrova, D., Garton, L., Gulia, M., & Haythornthwaite, C. (1996). Computer networks as social networks: Collaborative work, telework, and virtual community. *Annual Review of Sociology, 22*, 213-238.

Chapter VIII

Coordination and Control of Research Practice across Scientific Fields:
Implications for a Differentiated E-Science

Jenny Fry
Oxford Internet Institute, University of Oxford, UK

Abstract

This chapter speaks to the heterogeneity of research practices in science. It explores how cultural differences within and across disciplines shape the appropriation of e-science tools and infrastructures. Becher's (2001) anthropological perspective on academic disciplines and Whitley's (2000) organizational theory of scientific fields are used as a theoretical framework. The argument focuses on how differentials in the degree of interdependence between scientists and the level of uncertainty around research problems, objects, techniques and results affect the integration and coordination of work organization. The resulting cultural configuration has implications for mechanisms of control and consensus around the adoption of new technology. The chapter also highlights how appropriation can in turn shape the work organization and research practices of scientific communities.

Introduction

There are a number of problems with the current vision of a revolutionary science based upon technological promise. Evaluation of the uptake and use of e-science technologies is problematic due to the diversity of cultural identities across knowledge domains, which vary in terms of their social organization, intellectual goals, methods, techniques and tools and competence standards. There have been many failed projects and costly mistakes in the history of information and communication technology design, and while the successes are lauded, the possible reasons underpinning failures are inadequately evaluated. It is now obvious that given time all fields will not adopt e-science infrastructures into their work practices in the same way or to the same extent (Kling & McKim, 2000). We know that in some fields, such as genetics and biodiversity, there has been an increasing reliance on large-scale databases and information and communication technologies (ICTs) (Fujimura & Fortun, 1996; Lenoir, 1998; Bowker, 2000), which have become central to the research process, while in others digital tools and infrastructures are at best on the periphery of knowledge creation practices (Fry, 2004).

In the context of e-science, it has been pointed out that social science approaches often focus on the technically-driven hard sciences and consequently follow the flow of funding initiatives (Wouters, 2004). This means that outside of these fields we have a limited understanding of how research cultures shape the appropriation of e-science tools and infrastructures and how in turn the technology influences research practices and cultural identity. The development and use of e-science technology is emerging both inside and outside the boundaries of prestigious high-tech sciences such as genetics and high-energy physics. For example, in physical geography information scientists are working with practitioners to integrate multiple data sources into digital libraries to support higher education curricula (Borgman et al., 2005). In corpus-based linguistics, developments in designing and building mega corpora of language in use have burgeoned since the mid 1980s. Technical developments in corpus-based linguistics have generated new approaches in literary studies, amongst other fields, where efforts in the digitization of collections of texts and the development of text-analysis tools have been revolutionary in expanding the possibilities for the creation of new research problems (Hockey, 2000).

The purpose of this chapter is to demonstrate how two theories of the cultural organization of the sciences (Becher, 2001; Whitley, 2000) can be applied in the context of e-science ideologies and technologies to understand similarity and difference in patterns of appropriation. Systematic understanding of the mutual interaction between research culture and technology appropriation will be a valuable contribution to development initiatives. For example, it could be used to predict trends in uptake and use and identify potential areas for transdisciplinary collaboration and infrastructure building.

The explanatory framework provided by Becher (2001) and Whitley (2000) will be contextualized in the lived experience of scientific communities in three case-studies: high-energy physics, social/cultural geography and corpus-based linguistics. The main source of data for the three case-studies is in-depth interviews with thirty academics from across eighteen universities in the UK.

Differentiation across Science

Scientific knowledge is not a homogenous whole, but a patchwork of heterogeneous fields of collective research and pedagogical activities. As Haythornthwaite et al. (this volume) observe, specialist expertise is deeply rooted in disciplinary training and enculturation, to the extent that experts are so ingrained within the practices of their parent discipline that they are often unaware of how the ideologies, assumptions and language that they use in their day-to-day work shape the way they interact with the world and scientists from other disciplines. Haythornthwaite et al. argue that this has implications not only for the performance of interdisciplinary collaboration, but also for study of it.

The disciplinary view of science largely represents a market monopoly of certain types of knowledge embodied in university departments as training and employment units (Turner, 2000). Within the STS community it has been argued that knowledge is created at a finer level of granularity than the discipline (Knorr-Cetina, 1981). Articulating this fine-grained texture is problematic, however, because familiar terms such as "specialist field" are still perceived as representing an institutional perspective on knowledge creation (Weingart & Stehr, 2000; Knorr-Cetina, 1999). When we take a fine-grained perspective on scientific activity and knowledge creation, the cultural landscape becomes more complex than that which we take for granted when the notion of a scientific discipline is evoked.

This variegated landscape has been captured by sociology of science scholars with equally varied, but overlapping, concepts such as research fronts (Lenoir, 1997); specialisms (Becher, 2000); intellectual communities (Whitley, 2000); epistemic cultures (Knorr-Cetina, 1999; see also Wouters and Beaulieu this volume for a discussion of epistemic cultures); invisible colleges (Price, 1963); and discourse communities (Swales, 1998). These terms variously refer to the collectives of scientists that create and inhabit a bounded area of common research endeavour and to organizational entities discrete from disciplinary activities, in so far as researchers perceive themselves as advancing the research front, rather than the discipline. These areas are not necessarily localized in a single discipline, nor do they necessarily follow disciplinary boundaries (Lenoir, 1997). Indeed, they are often interdisciplinary, integrating ideas, methods and techniques from different disciplines to create new knowledge.

Despite criticism of the discipline as an appropriate unit of analysis for studying the work organization of the sciences (Knorr-Cetina, 1999; Fry, 2004b), there has been little by way of alternative frames of reference for systematic comparison of research cultures. This can perhaps be ascribed to the remaining importance of disciplines as the main providers of the scientific labor force and economies of practice. Lenoir (1997, p. 46) reminds us that disciplines are institutionalized formations for organizing schemes of perception, appreciation and action, and for their inculcation as tools of cognition and communication.

It is not the ambition of this chapter to resolve the debate around nomenclature, therefore the narrative will keep to the terms of "discipline" and "specialist field" as familiar "hooks" to differentiate between knowledge institutionalized in a market monopoly for pedagogy, training, employment and communication (the discipline) and knowledge as a non-monopolized uncertain research activity (the specialist field). I will mostly concern myself with those that work within these two realms simultaneously, for example, the focus will be upon academic scientists. These collectives of scientists will be referred to as disciplinary or specialist communities respectively; where granularity is not significant to the discussion, they will be generically referred to as scientific fields or communities.

Frameworks for Systematic Comparison

Taking an anthropological approach to understanding differentiation across disciplinary communities, Becher (1987) studied ten disciplines and devised a taxonomy of their cultural identities. His taxonomy is based upon earlier systems developed by Biglan (1973) and Kolb (1981) who were concerned with the extent to which knowledge structures can be classified as hard versus soft and pure versus applied. An important cultural dimension that Becher uses as a comparative device, and that influences the division of labor within disciplines, is whether knowledge is cumulative, which he describes as crystalline and increasingly specialized, or reiterative and revisionist, involving the same phenomena to be reexamined based on different perspectives. Another dimension is the type of outcomes produced, such as products and techniques in hard-applied disciplines in contrast to discovery and explanation in hard-pure disciplines. In the taxonomy, Becher emphasizes the close interrelationship that exists between the characteristics and structures of disciplines, and the attitudes, activities and cognitive styles of the communities of academics that occupy them, such as whether they are politically well organized or loosely structured. Becher aggregated the cultures he identified into four familiar broad disciplinary groupings, as shown in Table 1. The three case-studies constituting the main empirical examples in the chapter have been positioned in the taxonomy in italics.

Elegantly reminding us that scholarly communication is the very essence that binds together the social and the epistemic, Becher presents a compelling and

Table 1. Becher's taxonomy of disciplines represented in broad disciplinary groupings (adapted from Becher, 1987, p. 289)

Group	Knowledge	Culture
Physical Sciences, (e.g., physics, *high-energy physics*). "hard-pure"	Cumulative; atomistic (crystalline/tree-like); concerned with universals, quantities, simplification; resulting in discovery/explanation	Competitive, gregarious; politically well organized; high publication rate; task-oriented
Humanities (e.g., history, *corpus-based linguistics*) & Pure Social Sciences (e.g., anthropology, *social/cultural geography*) "soft-pure"	Reiterative; holistic (organic/river-like); concerned with particulars, qualities, complication; resulting in understanding/interpretation	Individualistic, pluralistic; loosely structured; low publication rate; person-oriented
Applied Sciences (e.g., mechanical engineering, *corpus-based linguistics*) "hard-applied"	Purposive, pragmatic (know-how via hard knowledge); concerned with mastery of physical environment; resulting in products and techniques	Entrepreneurial, cosmopolitan; dominated by professional values; patents substitutable for publications; role-oriented
Applied Social Sciences (e.g., education, *social/cultural geography*) "soft-applied"	Functional, utilitarian (know-how via soft knowledge); concerned with enhancement of [semi-] professional practice; resulting in protocols and procedures	Outward looking; uncertain in status; dominated by intellectual fashions; publication rates reduced by consultancies; power-oriented

insightfully rich description of the cultures of the disciplines and their styles of communication. It is, however, limited for understanding work organization and communication at the level of specialist communities. This is less problematic for a specialist field such as high-energy physics that occupies a central position within the parent discipline and has strong disciplinary boundaries. The culture of specialist fields that do not represent the disciplinary core, however, and function at the interstices between disciplines, such as social/cultural geography and corpus-based linguistics, are not captured in such broad categories. For example, geography is typically perceived as a physical science, due to the dominance of a "hard science" paradigm in the disciplinary core. However, geography is divided into two major subdisciplines: physical geography and human geography, which are then of course further divided into specialist fields such as climatology, geology and palaeontology along the physical dimension and social/cultural geography, economic geography and political geography along the human dimension. This can be problematic when disciplinary categories alone are used for evaluation and resource allocation.

The social/cultural geographers in the case study described their intellectual life and disciplinary experience as significantly different from colleagues in the physically-oriented fields that tend to be more prominent in academic geography departments. For example, while climatology draws on approaches and techniques in disciplines such as physics, chemistry and statistics (Borgman et al., 2005), social/cultural geography overlaps with anthropology and sociology, and falls between the humanities and applied social sciences within Becher's (1987) taxonomy. In seeking information for research, in addition to news sources and informal networks, social/cultural geographers use journals and monographs produced within anthropology and sociology more than in core geography publications, though they mainly disseminate their own work in geography journals. The type of primary data they use, their research methods and search strategies, typified by heavy reliance on browsing and printed sources, differ a great deal from those of climatologists, palaeontologists and geologists.

Two interrelated concepts identified by Whitley (2000) may be used to explore research practices across specialist fields, as they support iteration between the internal organizations of specialist communities and the wider scientific environment of the disciplines. Whitley claims that many of the major differences between fields can be explained in terms of two dimensions of scholarly work the:

1. degree of *mutual dependence* between researchers or fields in making competent and significant contributions to the body of knowledge; and the

2. degree of *task uncertainty* in producing and evaluating knowledge claims.

These two dimensions are interrelated, as they are further divided into four overlapping analytic elements that relate either to reputational control and coordination of research strategies and intellectual priorities, *strategic dependence* and *strategic uncertainty,* or to the coordination of competence standards, research techniques and task outcomes, *functional dependence* and *technical uncertainty*. Depending on the level of granularity, for example, discipline or specialist field, scientific communities may differ in their relative degree of "mutual dependence" and "task uncertainty," as some specialist fields may be more or less coordinated, integrated and standardized than their parent discipline. Whitley does not explicitly discuss his theory in the context of rapid expansion in ICTs, which has been an influential factor in shaping the scientific environment over the past two decades, therefore this narrative extends his theory by using it to predict differential in e-science practices.

Power Relations in Science

What unites disciplinary communities is that they operate within a social context embedded with long-standing practices and rituals such as peer review, recognition and reward, publication, recruitment and enculturation of students, organization of scientific labor into employment units and competition for funding and resources. Taking the analysis to an even coarser-grained level of granularity than the specialist field or discipline, that of the science system as a whole, the social position of a field within the wider scientific system is a major factor in determining its autonomy, coherence and direction and thus, its relative degree of "mutual dependence" and "task uncertainty." A major market mechanism of the discipline is production through supply and demand. In turn, market competition creates division of labour, which is held in place by power relations. During the industrial revolution, for example, changes in the structure and operation of science as a distinct institution producing distinct kinds of knowledge affected the internal organization of disciplines and their relative dependence upon one another. Thus disciplines depend upon each other to varying degrees for their own status and access to resources mediated by that status (Whitley, 2000).

The shifting relationship between disciplines in the wider scientific environment is important to bear in mind in the context of the emergent knowledge-based economy. Slaughter and Leslie (1999) argue that changes currently taking place in science are as great as the changes in higher education brought about by the industrial revolution. Globalization, they argue, is creating new structures, incentives and rewards for some aspects of academic careers and is simultaneously instituting constraints and disincentives for other aspects.

The economic structure of disciplines remains important to the development of specialist fields because, as Lenoir (1997) explains, the stabilization of diverse local practices occurs through their position in the broader context of an economy of practices. Therefore, Lenoir argues that disciplinary arrangements are crucial for organizing and stabilizing heterogeneity across fields. Lenoir (1997, p. 53) suggests that "the research front and discipline formation be treated as interrelated, not as cause and effect, but as mutual resource."

Important themes in Becher's (2001) taxonomy that are particularly pertinent in the context of understanding how, why and when specialist fields appropriate e-science tools and infrastructures are the notions of hierarchy in contrast to pluralism, and centralization in contrast to decentralization in the social organizations of disciplines. For example, case-study participants from both social/cultural geography and corpus-based linguistics inhabited interdisciplinary research areas and described themselves as practicing on the margins of their parent discipline. The social/cultural geographers, in particular, perceived their work to be peripheral within geography and felt that their research activities and outcomes were underrepresented with respect to central recognition, reward and employment structures. Collaborating more with anthropologists and sociologists than other geographers, the resulting intellectual and social isolation within the parent discipline imposed barriers to communication, intra-institutional collaboration, interpersonal recognition (Hagstrom, 1965) and career advancement due to a lack of infrastructure support.

Integration and Standardization of Research Procedures

The difference between being a specialist community able to influence technological development at the design phase, rather than being an end-user community at the appropriation phase, is partly determined by the nature of reputational control in a field, which has a significant influence upon patterns of communication. A high-degree of centralized reputational control, enacted through social formations such as invisible colleges (Crane, 1972; see Barjak this volume for a discussion of the influence of this phenomenon upon equality in information access), and cultural elites, will lead to domination of the major channels of communication in a discipline. Scientific communities also vary in the extent to which they adopt external criteria and standards for evaluating the significance and quality of research produced outside their own field. In some fields, reputations may be governed by norms derived from the dominant paradigm in the parent discipline and in others determined by indigenous decentralized goals and criteria.

A greater plurality of influences upon performance and significance criteria in a field, for example through decentralized reputational control, restricts the development of standardised research procedures and skills, and also limits the need and ability to coordinate and systematically compare task outcomes. This is typical of specialist fields where there is an increasing degree of "task uncertainty" and results are ambiguous and subject to a variety of conflicting interpretations. Researchers will deal with a variety of problems in a variety of ways and seek reputations for diverse contributions to broadly-conceived goals without having to demonstrate their specific connections to specialist colleagues' strategies. With limited collective control over what counts as an important problem or how problems should be formulated, research efforts and approaches may become fragmented (Whitley, 2000).

Technical procedures will be highly tacit, personal and fluid with an increased reliance upon direct and personal control of how research is conducted, local variations in work goals and processes and greater emphasis upon informal communication and coordination processes. Consequently, it is more difficult for communities inhabiting these fields to coordinate their work around a centralized ICT infrastructure. Compare, for example, high-energy physicists for whom Grid computing has been possible because the international community had previously developed a networked infrastructure interoperable across geographic and institutional boundaries, with corpus-based linguists who still face many challenges in the remote coordination of distributed collaboration due to radical interoperability and coordination issues, such as infrastructure inequalities across partners.

Table 2 illustrates the consequence of a decreasing or increasing degree of "task uncertainty" for the integration and standardization of research procedures (the three case studies constituting the main empirical examples in the chapter have been positioned in the taxonomy in italics).

"Internationalized" sciences such as high-energy physics tend to reflect the dominant characteristics of the most important national system in the field. Whitley (2000) argues that where one country in a particular field dominates research priorities and facilities to the extent that its scientists are able to set significance and performance criteria for the whole international system of research, scientists are primarily oriented to the reputational system of that one country and are primarily governed by its priorities and significance criteria. In fields where control over reputations, research priorities and facilities is diffused across national boundaries, such as corpus-based linguistics, national variations in intellectual priorities and styles of work are more important and noticeable. Scientists are more oriented to their national colleagues than to an international reputation system, which means that they are more likely to organize work around a national technological infrastructure, than a field-based one that transcends geographic and institutional boundaries.

Table 2. Influence of "task uncertainty" on the integration and standardization of research procedures (adapted from Whitley, 2000)

		Technical Uncertainty	
		Low	High
Strategic Uncertainty	Low	Considerable predictability, stability and visibility of task outcomes. Implications of results easy to draw and relatively uncontroversial. Problems and goals fairly clearly ordered, restricted and stable. (e.g., Chemistry, Physics, *High-energy physics*)	Limited technical control of empirical phenomena; results unstable and difficult to interpret. Implications of task outcomes subject to alternative views and difficult to coordinate. Problems and goals restricted, stable and tightly structured. (e.g., Economics, *Corpus-based linguistics*)
	High	As above, but problems and goals varied, unstable and not clearly ordered. (e.g., Biology since 1950, artificial intelligence, and engineering)	As above, but problems and goals varied, unstable and conflicting. (e.g., Post-1960 U.S. Sociology, post-1960 U.S. ecology, *Social/cultural geography*)

Whitley's concept of "strategic dependence" has direct consequences for the integration of goals and strategies between fields so that communities coordinate their approaches and problems. Where it is relatively limited, specialist communities are able to pursue distinct goals and objectives without concerning themselves overmuch with the significance of their reputations and their scientific status as determined by other fields or the parent discipline. Here, scientists do not seek to demonstrate how crucial their problems and concerns are for science as a whole or to influence colleagues in other areas. A greater degree of strategic dependence, as exemplified by high-energy physics, implies that the relative standing of scientific fields in science as a whole is more important to scientists and they engage in more intensive competition over their centrality to scientific ideals. Indeed, this is necessary in order to leverage political power in obtaining large-scale resources and funding for high-tech research.

Table 3 summarizes the consequences of either a decreasing or increasing degree of interdependence between scientists for the extent of coordination and control within scientific communities.

Table 3. Influence of "mutual dependence" on coordination and control in scientific fields (adapted from Whitley, 2000)

		Functional Dependence	
		Low	High
Strategic Dependence	Low	Weakly-bounded groups pursuing a variety of goals with a variety of procedures. Little coordination of results or problems. Low extent of division of labour across research sites. (e.g., Sociology, Management studies, *Social/cultural geography*)	Specialist groups pursuing differentiated goals with specific, standardized procedures. Considerable coordination of results and specialized topics, but little overall concern with hierarchy of goals. (e.g., Chemistry, Mathematics, *Corpus-based linguistics*)
	High	Strongly-bounded research schools pursuing distinct goals with separate procedures. High degree of coordination within schools, but little between them. Strong competition for domination of field. (e.g., German philosophy and psychology before 1933)	As above, but strong hierarchy of specialist goals. Competition over centrality of subfields to discipline. (e.g., Physics, *High-energy physics*)

Control Over the Means of Production and Communication

Whitley argues that the industrialization of research in the nineteenth century, with the introduction of laboratory research, enabled institute leaders to monopolize access to essential facilities. This brought about a major change in the degree of concentration of control over the means of production and communication and increased the degree of interdependence between scientists.

Concentration of control over access to the means of intellectual production and distribution is essential for fields such as high-energy physics that have a highly-specialized knowledge structure and require resource intensive experimental apparatus, leading to a high degree of "strategic dependence" in its social organization. Whitley argues that a consequence of an increasing degree of "strategic dependence" is that coordination goes beyond the technical level of integrating specialist contributions to a common goal. Coordination becomes a political activity that involves the organizations of programs and projects in terms of particular priorities and interests. The political imperative culminates in the setting of research agendas, allocation of resources and the influence over careers. In order for this process to be propagated it is necessary for a small number of employment units and research sites to dominate control over jobs, facilities, funds and channels of communication.

Whitley compares high-energy physics, which requires large-scale expensive apparatus, with many of the human and biological sciences, where research can be conducted with the more limited resources that are readily available in most universities and laboratories. In these fields, working materials are either available at low cost and easy to appropriate personally, as in the case of personal libraries, or else are a collective facility controlled by larger organizations than individual departments, such as university libraries, so that particular reputational elites cannot dominate access to them and therefore there is less need to compete for control over resources (Whitley, 2000). Whitley describes most of the modern sciences in the UK as having some important resources centrally controlled, with stratified allocation to research sites, and each employment unit remaining relatively independent and able to have some autonomous control over goals and strategies.

In some fields, publishers play a central role as gatekeepers to scientific knowledge and can be powerful actors in delineating the boundaries of a discipline or specialist field through their publication policy. The policies of publishers are then reflected on digital networks through a myriad of representational devices (Latour, 1987) such as specialized digital libraries and on-line catalogues. In fields where textual practices tend to focus on production of monographs, concentration of control over the means of communication is further decentralized from the control of dominant groups. Instead, publishers are in a position where they can define what topics and approaches are valid and competent according to commercial goals. For example, the social/cultural geographers described a number of reorientations of the field that had warranted a revision of how they labelled their own research in order to align it with the mission of the main publishers in the field. In the past decade in human geography there has been something of a paradigm shift, which the social/cultural geographers described as "the cultural turn."

There are fashions as well. If you're asking what's changed in my research field recently there's a thing called the 'cultural turn'. [It's] the dominant trend where all human geography is now about cultural questions, rather than economic or political whatever, I think that's a fashion. (Social/cultural geographer, 1998)

One group, in particular, had been having difficulty publishing their work because the heightened interest in social and cultural work meant that it was becoming increasingly difficult to publish work in other fields of human geography, as many book and journal publishers responded to the intellectual shift. This is where groups outside of the field, such as a lay audience can be influential. To a certain extent, the rise of the university press was a way to alleviate the risk of book publishers dominating the communication system and influencing academic careers (Swales, 1998). In certain fields, however, this obviously does not have a significant impact on diminishing the effect of the economic imperative of publishers. Looking at how fields are differentially represented on publicly available digital networks (Fry, forthcoming) it is apparent that in those fields where book publishing is the dominant mode of communication and thus reputation building, publishers have a great deal of control over how those fields are represented and when, how and who can access research outcomes. As Barjak (this volume) concludes, *"it is the hierarchy and power distribution in science, not the technology, which determines access to information."*

Stabilization through Technology

As we have seen, variation in the degree of "mutual dependence" between scientific fields reflects their extent of autonomy or interdependence. The most prominent manifestation of "functional dependence" is the use of technical procedures and instruments from neighboring fields. If the relative degree of "functional dependence" between two or more fields is high, scientists will be more reliant on the results and products from one another and use each other's ideas and procedures fairly frequently. The appropriation of e-science tools and infrastructures has considerable implications for this aspect of work organization within fields. Vann and Bowker (this volume), for example, discuss the labour process in interdisciplinary collaboration and how narrative scripts of technological prospects redraw the division of labour across fields, with the goal of enhancing the study of phenomena by multiple disciplinary approaches.

Whitley (2000) argues that the more a hierarchy of goals and problems is accepted in the sciences, the more scientists orient their strategies and research programs to higher-ranked goals and seek to acquire reputations for contributions to them. Scientific fields become more coordinated around some central objectives and conceptions of scientific knowledge so that strategies are more interconnected across fields. For example, the transfer of techniques and instruments from physics to chemistry and biology implies that the approach and goals of physics that they manifest are also significant for chemistry and biology.

Trading zones (Galison, 1997), where instrument makers, theorists and experimentalists meet around large technical infrastructures, can be fertile territory for interdisciplinary innovation. Despite increased funding for interdisciplinary collaboration by the UK research councils, there remains a discrepancy between the pursuance of interdisciplinary research at the level of specialist communities and support in terms of institutional arrangements. For example, in the UK there has been controversy over the Research Assessment Exercise (RAE) as a disincentive to interdisciplinary work, and it is thought to be difficult for postgraduates with interdisciplinary degrees to get jobs (Turner, 2000). This leads to a differential in the perceived valency of cross-disciplinary fertilization through technology.

Computer-based techniques, approaches and tools in corpus-based linguistics, for example, have led to differential innovative practices both within linguistics and its disciplinary neighbor, literary studies. In the mid-1990s the field of contrastive linguistics and cross-linguistics research reemerged largely due to technological developments in corpus-based linguistics (Salkie, 2004). Through the assimilation of digital tools such as taggers and parsers, together with increased computational power that enabled the building of large accurate multilingual corpora, the field of contrastive linguistics reformulated its research problems, techniques and audience. Meanwhile, fields in the neighboring discipline of literary studies have also assimilated corpus-based tools and techniques. Parser techniques originally designed for analysis of texts at the linguistic level, for example, syntax or morphology, were adapted for the literary analysis of texts by researching changes in narrative styles of authors over time or researching the use of names in a particular novel (Van Dalen-Oskam & Van Zundert, 2004).

At the same time there has been a difference in the perceived valency of these tools and techniques between the field of corpus-based linguistics and its parent discipline, linguistics. Since the reemergence of corpus-based linguistics itself in the 1950s, stimulated by the first computer-based language corpus, the community has been entrenched in an ongoing debate within the parent discipline about the validity of its probabilistic approaches and methods for the development of linguistic theory. The core of the debate revolves around different conceptions of what counts as legitimate data, such as the use of "empirical" data observable in a corpus, or the use of "subjective" data based on native speakers' intuition of the language under study.

The main objection to the validity of corpora is that they are perceived as insufficiently representative of a particular speech community to make a contribution to linguistic theory. Elements such as the size of a corpus, the number of words, source material, for example, spontaneous speech, parliamentary debates, television broadcasts, newspapers, etc., and scope of coverage, for example, age, race, class and dialect, have often led corpora to be criticized and dismissed as too skewed a representation. The second objection to approaches within corpus-based linguistics is based on the communities' use of computational linguistics, at the core of which are probabilistic techniques. At times, the tone of this debate has been somewhat polemic, often spurring a flurry of articles in the scientific journals (see Borsley & Ingham, 2002; Stubbs, 2002; and Borsley & Ingham, 2003, for an example of this type of exchange). The struggle of corpus-based linguistics to gain status as a valid approach within linguistics (Sampson, 2003) has meant that it has a relatively low degree of "strategic dependence," which acts as a barrier to scaling up tools and resources into an interoperable field-level infrastructure.

Establishing technical standards for interdisciplinary collaboration in fields where there is a low degree of "strategic dependence" can be problematic, because different specialist communities bring different resources, skills and goals to the interaction. For example, speech technologists come from a culture akin to Becher's (1987) hard-applied knowledge structure, whereas the sociolinguists better fit his description of a soft-applied knowledge structure. We can observe these tensions in how technical standards are evolving in the field. Atwell, et al. (2000) conducted a survey of the use of parsing standards within corpus-based linguistics and found that there was much local variation in the use of parsing schemes:

The rather disheartening conclusion we can draw from these observations is that it is difficult, if not impossible to map between all the [parsing] schemes... No single standard can be applied to all parsing projects. Even the presumed lowest common denominator, bracketing, is rejected by some corpus linguists... . The guiding factor in what is included in a parsing scheme appears to be the author's theoretical persuasion or the application they have in mind. (Atwell et al., 2000)

Though technical standards are necessary for interoperability, there has been a resistance to data standards in many humanities fields because they are perceived as necessitating the standardization of research objects and imposing a normative practice. In general this conflicts with the pluralistic and reiterative nature of the humanities (Becher, 2001), where there is a low-degree of "functional dependence" and the values and goals incorporated in the technologies of one field are less likely to be shared by another. Therefore, the existing set of social relations resists reorganization by new technology depending on the status of those proposing an innovation and the value system of the target community (Thomas in Walsh & Bayma, 1996). In contrast to the large-scale centralized e-science infrastructures of high-energy physics and genetics, what we observe across the humanities are numerous small-scale local developments taking place.

According to Hockey (2000), the humanities computing community has been leading broad developments in linguistics, literary studies and history. The leaders of this community have developed the Text Encoding Initiative (TEI) (Burnard, 1988), which has become the dominant standard for digitization projects in humanities computing and presumes a particular philosophical approach to working with and analyzing texts. For example, Hockey (2000, p. 45), a proponent of the Text Encoding Initiative (TEI), writes, "[m]ore than anything, by encouraging a thorough analysis of the features in a text, the TEI ensures that encoders think seriously about the material they are encoding." As such, the standard plays a central role in legitimating research. For example, references to the TEI are prominently positioned in the literature and web sites of many national corpora projects, such as the British National Corpus and the Corpus of Spoken Dutch, even though communities, such as speech technologists, involved in building and working with these corpora have a high technical proficiency and have developed alternative standards. To a certain extent, citing the dominant standard in the published literature and project documentation is a political act, perhaps necessary in order to receive further funding. To be aligned with the dominant standard is to be perceived as competent. Technical standards, then, become part of the political economy of a discipline or field. They become a social means for controlling the production and communication of knowledge and stabilizing the research object.

Conclusion

This chapter has offered a systematic explanation of why it is not *"just a matter of time"* (Kling & McKim, 2000) until all fields adopt e-science tools and infrastructures into their work practices in the same way or to the same extent.

An increasing degree of "strategic dependence" increases the likelihood that a field will be governed by a single reputational group that will control access to the means of production and communication. This means that scientific communities will seek to align their goals, approaches and techniques through integration and are likely to appropriate technology to support this. At the same time the motivation and ability of scientific communities to coordinate and control the standardization of research objects varies across fields depending on the relative degree of "functional dependence." Where there is an increasing degree of "functional dependence" between fields this can be expected to reduce the strength of intellectual and organizational boundaries and encourage the generation of transdisciplinary techniques and procedures. It is in these areas of research then that we might expect to see interdisciplinary innovation through the adoption of e-science tools and techniques. An increasing degree of "task uncertainty," however, negates the ability to integrate and coordinate research goals, techniques and tools.

Consequently, specialist communities that create and inhabit fields with a relatively low degree of "mutual dependence" coupled with a relatively high degree of "task uncertainty" are less able to coordinate around and create e-science initiatives on the scale demonstrated by the UK high-energy physics community when PPARC (Particle Physics and Astronomy Research Council), the UK national funding agency for high-energy physics, received 26 million pounds from the British Government in 2000 to support development of the technological infrastructure necessary for the realization of Large Hadron Collider (LHC) facilities. One physicist explained that despite the competition between each experiment, the community stood shoulder-to-shoulder when it came to policy-making and developing an external representation of the community. The high-energy physics community is a powerful lobby in acquiring large sums and persuading peers of the centrality of the field to science, because culturally there is a high degree of "mutual dependence" coupled with a low degree of "task uncertainty."

This cultural configuration is not confined to the technically-driven hard sciences, and should not be extrapolated as such. For example, the humanities field of argumentation theory, an interdisciplinary formation from philosophy, linguistics and communication science, also has a high degree of "strategic dependence." Though the resources at stake are minimal in terms of financial costs and labour force compared with high-energy physics, the lead group in the field holds a political position of power where they are setting the research agenda for the field, controlling the main channels of communication, have a great deal of reputational control and therefore are setting the norms for the field (Fry, forthcoming). Becher (2001) describes the formation of strong group behavioral standards in highly cohesive scientific communities where there is a strong paradigm present. Deviance from the group norm in tightly-knit communities incurs social penalties such as exclusion from the communication system. We can hypothesize then that until the dominant reputational group in argumentation theory perceives e-science practices to have valency for the field the rest of the community will resist e-science tools and infrastructures.

Comparison of the three main case study fields in Table 4 provides a snapshot of the constitutive relations between cultural identity along the dimensions of "mutual dependence" and "task uncertainty," and the coordination and control of work practices through digital networks.

Given that the relative status and interdependence between intellectual fields changes over time in relation to shifts in societal priorities, current power relations may alter in the context of an emergent knowledge-based economy and e-science infrastructures. This in turn will bring about change in patterns of

Table 4. Relationship between cultural identity of fields and coordination and control through technology (adapted from Fry, 2006).

Field	High-energy physics	Corpus-based linguistics	Social/cultural geography
Culture	High degree of "mutual dependence"/low degree of "task uncertainty.."	Moderate degree of "mutual dependence"/moderate degree "task uncertainty."	Low degree of "mutual dependence"/high degree of "task uncertainty."
Coordination	Centralized coordination of research problems, strategies and techniques at the level of the international community.	Coordination of research problems, strategies and techniques subject to local interpretation, with distinct national approaches.	Decentralized coordination of research problems, strategies and techniques allows for intellectual pluralism, but restriction to core disciplinary methods due to lack of internally produced significant criteria for assessing imported methods.
Collaboration	High people-to-problem ratio, goals tightly coordinated and integrated through a high degree of specialization contiguous with the parent discipline.	Uneven people-to-problem ratio (e.g., dominance of English as research object), goals coordinated and integrated through interdisciplinary projects with limited lifecycle.	Low- people-to-problem ratio, decentralized coordination of goals subject to local variation in division of labor.
Appropriation of ICTs	Community led digital infrastructure at core of knowledge creation and collaboration practices.	Community led digital tools, rather than infrastructure, at core of knowledge creation and collaboration practices.	Institutionally led digital resources and infrastructures at the periphery of knowledge creation and collaboration practices.

interdependence and the integration of research goals and techniques across fields.

References

Atwell, E., Demetriou, G., Hughes, J., Schiffrin, A., Souter, C., & Wilcock, S. (2000). A comparative evaluation of modern English corpus grammatical annotation schemes. *ICAME Journal, 24,* 7-23.

Becher, T. (1987). The disciplinary shaping of the profession. In B. R. Clark (Ed.), *The academic profession* (pp. 271-301). Berkeley: University of California Press.

Becher, T. (2001). *Academic tribes and territories: Intellectual inquiry and the culture of disciplines* (2nd ed.). Buckingham: SHRE & Open University Press.

Biglan, A. (1973). The characteristics of subject matter in different academic areas. *Journal of Applied Psychology, 57*(3), 195-203.

Borgman, C. L., Smart, L. J., Millwood, K. A., Finley, J. R., Champeny, L., Gilliland, A. J., & Leazer, G. H. (2005). Comparing faculty information seeking in teaching and research: Implications for the design of digital libraries. *Journal of the American Society for Information Science & Technology, 56*(6), 636-657.

Borsley, R. D., & Ingham, R. (2002). Grow your own linguistics? On some applied linguists' views of the subject. *Lingua, 112,* 1-6.

Borsley, R. D., & Ingham, R. (2003). More on "some applied linguists": A response to Stubbs. *Lingua, 113,* 193-196.

Bowker, G. (2000). Biodiversity datadiversity. *Social Studies of Science, 30*(5), 643-84.

Burnard, L. (1988). Report of workshop on text encoding guidelines. *Literary and Linguistic Computing, 3,* 131-133.

Crane, D. (1972). *Invisible colleges: Diffusion of knowledge in scientific communities.* London: The University of Chicago Press.

Fry, J. (2004). The cultural shaping of ICTs within academic fields: Corpus-based linguistics as a case study. *Literary and Linguistic Computing, 19*(3), 303-319.

Fry, J. (2006). Scholarly research and information practices: A domain analytic approach. *Information Processing and Management, 42*(1), 299-316. Retrieved November 24, 2004, from http://www.sciencedirect.com/science/journal/03064573

Fry, J. (forthcoming). Understanding disciplinary differential on the web. *Cybermetrics.*

Fujimura, J., & Fortun, M. (1996). Representing nature in bytes: The construction of molecular genetic sequence databases. In L. Nader (Ed.), *Naked science: Anthropological inquiry into boundaries, power, and knowledge* (pp. 160-173). New York: Routledge.

Galison, P. (1997). *Image and logic: A material culture of microphysics.* Chicago: The University of Chicago Press.

Garfield, E. (1964). Science citation index: A new dimension in indexing. *Science, 144*(3619), 649-654.

Garvey, W. D. (1979). *Communication: The essence of science.* Oxford: Pergamon Press.

Granstrand, O. (1999). *The economics and management of intellectual property: Towards intellectual capitalism*. Cheltenham, UK: Edward Elgar.

Hagstrom, W. (1965). *The scientific community*. London: Basic Books.

Hockey, S. (2000). *Electronic texts in the humanities: Principles and practice*. Oxford: Oxford University Press.

Kling, R. (1991). Social analyses of computing. In C. Dunlop & R. Kling (Eds.), *Computerization and controversy* (pp.150-166). Boston: Academic Press.

Kling, R., & McKim, G. (2000). Not just a matter of time: Field differences and the shaping of electronic media in supporting scientific communication. *Journal of the American Society for Information Science, 51*(14), 1306-1320.

Kling, R., Spector, L., & Mckim, G. (2002). Locally controlled scholarly publishing via the internet: The guild model. *The Journal of Electronic Publishing, 8*(1). Retrieved March 4, 2004, from http://www.press.umich.edu/jep/08-01/kling.html

Knorr-Cetina, K.(1981). *The manufacture of knowledge: An essay on the constructivist and contextual nature of science*. Oxford: Pergamon.

Knorr-Cetina, K. (1999). *Epistemic cultures: How the sciences make knowledge*. Cambridge, MA: Harvard University Press.

Kolb, D. A. (1981). Learning styles and disciplinary differences. In A. Chickering (Ed.), *The modern American college* (pp. 232-255). San Francisco: Jossey Bass.

Latour, B. (1987). *Science in action*. Cambridge, MA: Harvard University Press.

Law, J. (1973). The development of specialities in science: The case of x-ray protein crystallography. *Science Studies, 3*, 275-303.

Lenoir, T. (1997). *Instituting science: The cultural production of scientific disciplines*. Stanford, CA: Stanford University Press.

Lenoir, T. (1998). Shaping biomedicine as an information science. *Proceedings of the 1998 Conference on the History and Heritage of Science Information Systems*.

Leydesdorff, L. (2000). *A sociological theory of communication: The self-organization of the knowledge-based society*. Parkland, FL: Universal Publishers.

Leydesdorff, L. (2005). The knowledge-based economy and the triple helix model. In W. Dolfsma & L. Soete (Eds.), *Reading the dynamics of a knowledge economy*. Cheltenham: Edward Elgar, (forthcoming).

Leydesdorff, L., & H. Etzkowitz (1998). The triple helix as a model for innovation studies. *Science and Public Policy, 25*(3), 195-203.

NSF (2002). *Science & Engineering Indicators—2002 (NSB 02-1),* Chapter 5 academic research and development.

Price, D. J. (1963). *Little science, big science.* New York: Columbia University Press.

Salkie, R. (Personal communication April 7, 2004).

Sampson, G. (2003). Are we nearly there yet, mum? *Corpus Linguistics 2003.* Retrieved March 4, 2004, from http://www.grsampson.net/Aawn.html

Slaughter, S., & Leslie, L. (1999). *Academic capitalism: Politics, policies, and the entrepreneurial university.* London: The Johns Hopkins University Press.

Stubbs, M. (2002). On text and corpus analysis: A reply to Borsley and Ingham. *Lingua, 112,* 7-11.

Swales, J. M. (1998). *Other floors, other voices: A textography of a small university building.* London: Lawrence Erlbaum Associates, Publishers.

Turner, S. (2000). What are disciplines? And how is interdisciplinarity different? In P. Weingart & N. Stehr (Eds.), *Practising interdisciplinarity* (pp. 46-65). Toronto: University of Toronto Press.

Van Dalen-Oskam, K., & Van Zundert, J. (2004). Modeling features of characters: Some digital ways to look at names in literary texts. *Literary and Linguistic Computing, 19*(3), 289-301.

Walsh, J. P., & Bayma, T. (1996). Computer networks and scientific work. *Social Studies of Science, 26*(4), 661-703.

Weingart, P., & Stehr, N. (Eds.). (2000). *Practising interdisciplinarity.* London: University of Toronto Press.

Whitley, R. (2000). *The intellectual and social organization of the sciences* (2nd ed.). Oxford: Clarendon Press.

Wouters, P. (2004). *The virtual knowledge studio for the humanities and social sciences @ the Royal Netherlands Academy of Arts and Sciences.* Amsterdam: the Royal Netherlands Academy of Arts and Sciences. Retrieved July 27, 2005, from http://www.knaw.nl/cfdata/news/latestnews_detail.cfm?nieuws__id=272

Section III

Prospects for Transformation

Chapter IX

Cyberinfrastructure for Next Generation Scholarly Publishing

Michael Nentwich
Austrian Academy of Sciences, Austria

Abstract

This chapter deals with the future of scholarly publications as a key element of the knowledge production process of science and research. Publications are both at the input and the output side of knowledge creation and an important means of communication among scientists. In the age of cyberscience, or e-science, the publishing system is changing rapidly and we expect more fundamental changes to come as soon as most scholarly publishing has gone online and researchers have started to explore the new opportunities. A new kind of infrastructure is emerging that will add new actors to the traditional ones and potentially adds new functions and mechanisms. The chapter outlines the status quo and new technological as well as organizational options for scholarly publishing and develops a scenario of the next generation academic publishing system. It concludes with practical recommendations for designing the scholarly e-publishing cyberinfrastructure of the future.

Introduction and Approach

Scholarly publications are a key element of the knowledge production process of science and research. They are both at the input and the output side of knowledge creation and they are an important means of communication among scientists. Hence, science and technology studies (STS) were and are interested in how publishing functions and develops. In the age of cyberscience—a notion that rather stresses the "soft" communicative aspects of the changing working environment (Nentwich, 2003)[1]—or e-science—focusing on global collaboration via the so-called Grid technology in the "hard" sciences (e.g., Hey & Trefethen, 2002)—the publishing system is changing rapidly. We expect more fundamental changes to come as soon as most scholarly publishing has gone online and researchers have started to explore the new opportunities, such as multimedia. A new kind of infrastructure—independent of the Grid—is emerging that adds new actors, such as the individual researcher who "self-publishes," to the traditional ones, such as commercial publishing houses, and will change their roles. Furthermore, this infrastructure potentially adds new functions, such as quality labelling, or new mechanisms, such as self-archiving or use-tracking.

This chapter is based on an encompassing technology assessment (TA) study (Nentwich, 2003) whose main research question was how, specifically, will technological developments change the ways research is done. To answer such a broad question is no easy task, as cyberscience is a moving target and an elusive subject since research done in the area is fragmented and often unsystematic. In particular, STS is often about either "S" that is science (e.g., how scientists arrive at results) or about "T" that is technology (e.g., how society reacts to a new technology or how the latter is shaped by the former). Only seldom it is about both at the same time: about technology (use) in science. TA is a special, practically-oriented branch of STS. Cyberscience has many characteristics of an ideal subject for technology assessment: It is about emerging technologies whose impacts are already partly visible in the present and which have the potential of widespread application. It calls for an encompassing study because the various aspects are strongly interrelated. The topic needs to be treated in an interdisciplinary manner, as the impacts are in the political, cultural, legal, economic and social sphere. Last, but not least, it is about an ongoing development reaching into the future, so it makes sense to look at it not only from an analyst's perspective, but also with a view to eventually formulating policy recommendations. Furthermore, this study and hence this chapter is informed by diffusion research, which is interested in how technological (or other) innovations are implemented (or not) in a social system, including organizations (Rogers, 1995). At the heart of this type of analysis are the innovation-decision process and the conditions for a successful innovation process. Hence, this stream of research contributes to an explanation of the status quo in any given situation of a diffusion process. The main focus is on individual actors (users), and their behavior and attitudes. While most diffusion research analyses past innovation processes without dealing with long-term consequences, some also includes trend and impact analysis. However, as there was no concept specific enough to satisfactorily grasp the phenomenon of cyberscience, a conceptual framework specific for the study had to be developed. This framework describes the scholarly communication system; the factors that have influenced the evolution from the traditional, non-ICT-based communication system to the present interim status quo, and which are deemed to play a role in the future path to cyberscience as well; and it links the observed changes to an assessment of the impacts on the scholarly communication system (Nentwich, 2005).

In this chapter, I shall first outline the status quo and new technological as well as organizational options for scholarly publishing. Second, I shall develop a scenario of the next generation academic publishing system, considering five essential dimensions—print, quality, economics, law and collaboration. Finally, I shall conclude with a few recommendations that follow from the preceding analysis.

Scholarly Publishing: Status Quo and New Options

There can be no doubt that electronic formats of publishing have gained ground in the academic world. In some fields, such as physics, economics and mathematics, e-preprint servers are a very important first step in the publication chain. Even in areas where there was no preprint and working paper culture before the advent of the internet (such as political science), e-print series are becoming widespread. The rise of peer-reviewed e-journals from a handful in the early 1990s to several thousands today[2] is another sign of a changing publishing environment. Today, there are only few traditional academic journals left that have not yet a parallel publication on the web. Many of these electronic (versions of) journals provide new services and enhancements for the benefit of the user, such as online prepublication, multimedia content, full text search, alerting service, cross-citation linking, etc. Submitting manuscripts as well as handling the refereeing process is largely done electronically, in many cases via e-mail, often using sophisticated web tools. To a much lesser extent than working papers and journals, books are also published online, at least partly or in parallel. While dissertations often do not appear in print any more, other publishers specialize in print-on-demand, meaning that they print and bind a volume only when a customer is not prepared to print it himself. In sum, the traditional formats of publication all have gone online already at least to some degree.

Digitization also opens up new ways of representing data, text and knowledge and many academics are already tinkering with these new options. There are very interesting and innovative publishing formats, which can hardly be compared with present academic publishing (for an overview, see Nentwich, 2003, 327ff.). Some of them stretch the concept of the academic journal. For instance, there are "living reviews" that publish regularly updated state-of-the-art reports, and there are interactive multimedia journals offering online debates on their articles together with innovative presentation formats. Other publication formats are completely new as they rely on the web completely, such as link collections, FAQs (collections of frequently asked questions and answers for academic purposes) and self-(pre-)publishing. Increasingly, scholars and scientists make their research results available in shared databases and set up electronic archives of primary sources or study protocols. Furthermore, certain contributions to academic newsgroups and discussion lists—a widespread and new type of academic communication—may be counted as a new type of publication, too. This may be the case if two conditions are met: first, if the posting is more than an announcement or question, e.g., an elaborate answer or comment; and second, if the postings are archived. Stevan Harnad coined the label "scholarly skywriting" (1990) for this.

Electronic publishing in academia is in constant flux, not only quantitatively, but also in the sense that new types are developed and tested. For instance, what we may call "crossover publications" emerge: As length of academic publications is no longer a principled problem in the e-publishing world and as modularization would enable layered publications with multiple access and varying depth according to reader or reading purpose, the boundaries between the various formats of publications may become permeable. The crossovers would fall in neither traditional category, they might be read as a short journal article and simultaneously as a richly documented research report, as a research abstract and as a book-long argument; some are a mix between "special issues of e-journals" and "edited e-volumes"[3]; and "e-readers" would be something in between a "distributed book" and a "review article" as they comment on papers published elsewhere and place them in a new electronic context.

Another area of vibrant development is hypertext (Nentwich, 2003, 257ff.). While most of the earlier electronic publication formats firmly stand in the tradition of linear text, newer concepts place emphasis on "layered" and "networked" knowledge presentation. Indeed, hypertextuality was always present in academic writing (footnotes, indices, marginal text, etc.). However, cyberscience has the potential to go well beyond the status quo of academic knowledge representation due to the electronic environment of networked computers and databases. As mentioned above, already today a number of features of electronic publications are due to hypertext: linking between different parts of a text, links to external resources, etc. Modularity, that is the breaking of the linear text into smaller pieces and rearranging these modules with hyperlinks to a network of text, enables completely new forms of knowledge presentation and science communication. While we have already today a web of linear articles, new publishing models include layered e-publications, field-wide thematic hyperbases, distributed hyperbooks and knowledge bases containing the consolidated state-of-the-art of a field.

Next Generation Scholarly Publishing

Starting from the above overview of the status quo of academic publishing and our glimpse of the possible future, our next step is to develop a reasonable scenario of next generation scholarly publishing. The following aspects need to be considered: first, the fate of print publications; second, the type and organization of quality control in cyberscience; third, the changing economic and, fourth, legal environments of academic publishing; fifth, the compatibility of the emerging system with the increasingly collaborative cultures in the various fields.

The Future of Print Publications

It may well be that e-publishing supersedes old print publishing in academia, leaving some niches only for a restricted print market. This will, however, not be the case in the foreseeable future. We have to differentiate. First, the market for academic publications should not be confused with the one for trade books. Second, and even more importantly, there are a variety of publication formats within academia, which have to be treated differently. Third, the various disciplines' reaction to the challenge and option of e-publishing will not be identical. Many factors impact differently depending on the field-specific case. On a general level, I found that the technical aspects would not be crucial. More important is the destiny of the much higher prestige of paper print as opposed to digital media. This, in turn, depends on a variety of factors. Their power is difficult to predict and so is, in the end, the final result of the ongoing development.

Based on my careful analysis of the various factors and arguments in favor and against (for details see Nentwich, 2003, 349ff.), I am nonetheless in a position to venture a prognosis for the medium run, that is for the next five to ten years.

Table 1. The print to electronic scenario matrix: Predicting publishing formats in the medium term (Nentwich, 2003, 357)

		Publishing media		
		P-only	**PoD**	**Hybrid**	**E-only**
	Monograph	x	x	x	x
	Thesis		x		x
	Edited volume				x
	Reader				x
Academic publishing formats	**Reference work**			x	x
	Proceedings				x
	Text book		x	x	x
	Edition			x	x
	Journal			x	x
	Review journal				x
	Working paper				x
	Reports		x	x	x

Table 1 distinguishes between, on the one hand, the different academic publication formats and between the three alternative technologies (media) plus the base scenario of print-only publishing, on the other hand. The four options are not mutually exclusive. The suffix "-only" indicates that the publication would not have two different versions, but only one. Similarly, "Hybrid" is an independent variant, namely that the publication comes in two (perhaps different) versions. Hence, if for one publishing format there is one "x" for "Hybrid" and one for "E-only," this means that there may be some publications in this category with a double face (print and digital) and others with only a digital face.

A few words interpreting the matrix: In the medium run, the *monograph* will not go entirely digital. The publishing medium for the scholarly monograph will encompass print, print-on-demand (PoD), digital and hybrid forms. The pure e-book will be the exception for still some time. Only in the longer run, might there be a more important segment of books published as e-only books. They will be not comparable to the books we know today, but they will rather be "hyperbooks." My expectation is that the print-only *journal* will vanish very soon and be replaced by hybrid print and electronic journals practically everywhere. The majority of experts included in the Swiss Delphi study (Keller, 2001) expected the print-version to vanish by 2005 already. This was too early, but I would not count on a long life for the print-only journal. Even those hybrid forms will lose their importance in the medium run. Within the next couple of years, journal articles will be published e-only. Review journals will soon not be published other than as e-journals. *Article collections* in book form will vanish as printed texts in the not so distant future. Edited books will be replaced by e-journal-like publications, readers by commented link collections and proceedings by data-base-like web sites. *Reference works* will exist in hybrid forms for still some time, but probably go online completely in the medium run. *Theses* and *reports* will be on offer both electronically and through print-on-demand. *Working papers* will be distributed only in digital format. *All* academic publishing formats lend themselves for e-only publishing, the majority of them, however, not as the only option.

Quality Control in Cyberscience

Mechanisms to ensure the quality of academic publications are an essential part of the academic system. There can be no doubt that the excellence of these mechanisms is a prime yardstick for the future publishing infrastructure. Cyberscience shows a considerable potential to change the system of quality control, but a number of hurdles have emerged that make adjustments of important traditions necessary. First, academic quality is, in principle, medium-independent. Whatever the medium, quality may be low or high, depending on the quality control system applied. Second, cyberscience brings about new forms of quality control, which have the potential to improve the traditional systems, if carefully implemented (in particular as add-ons and only partial replacements of traditional forms). In particular, the new forms of *ex post* control, such as rating—readers give marks to what they read—and use-tracking—which items groups or readers have read, used or quoted is recorded and aggregated results are fed back to readers—are not feasible in the paper-based world of publishing. In some respect, they may revolutionize scientometrics, in particular when it comes to evaluating types of citations and discriminating between them. Based on this new set of information about a particular published item, sophisticated systems of quality labelling and filtering may be set up (Nentwich, 1999). Third, working-paper archives and journals will probably offer both *ex ante* and *ex post* open peer commenting. This may turn scholarly publishing into a much more communicative process than hitherto.

Since interactive (as opposed to one-way) communication plays a significant role in most of the new formats of quality control, the single most important factor seems to be time, namely the time needed for such enhanced communication. It is difficult to predict—and largely independent from cyberscience develop-ments—whether publishing will continue to be ruled by the requirement to publish as much as possible or whether quality becomes more important. In the first case, there will only be very limited or no additional time available for more commu-nication. Hence, the move to the internet of the traditional system of quality control is likely to continue, but a qualitative metamorphosis would be less probable. Here, cyberscience's main impact would be that it speeds everything up. In the second case, by contrast, there would be more time available for communication, for instance in the form of skywriting and commenting. Here, the potentials could be exploited more fully and cyberscience may indeed contribute to and accelerate this change of scholarly work. Which of the two scenarios will be realized, depends largely on the publishing traditions of the various fields and how they develop. The further evolution of quality control systems in the various disciplines would probably not be synchronous.

Economic Environment

The scholarly publication system of the future will be influenced not least by economic considerations and constraints. Given the new electronic publishing opportunities and the financial crisis of the present, largely commodified system with strong commercial publishers, a system change is likely. Most probably, there will be a mixed system—partly commodified, partly de-commodified (Nentwich, 2001). The future of the evolution of the system of formal scholarly publication has not yet fully taken shape, but will most likely be characterized by a strong de-commodified core with some niches for commercial publishers. The core will probably be a publishing system run by scholarly associations and academic institutions (universities, libraries, etc.) where scholars upload their e-"prints" on central servers ("e-pre-prints"). These papers may undergo subsequent quality control, either by being submitted to the editorial boards of e-journals or by some other innovative rating process organized by scholarly associations. Although this system change is gradual and evolutionary, once completed the outcome will be fundamentally different from the status quo ante. If my analysis holds, it will be more than the search for a new balance of power between the major players (commercial publishers, university presses, libraries, universities, scholarly associations), but instead a considerable upgrading of the role of all players except commercial publishers.

Among the main reasons for this development are those at the economic level—the crisis of the present system and how the academic community is about to react to this challenge—and those at the technical level—the new opportunities of the ICT, which have already taken over many of the functions performed by the private sector. In addition, ideational developments herald the future system—the growing conviction that almost free access to scholarly work on the internet is a good idea, and the open access perspective which could slowly develop normative power and de-legitimize present arrangements. Recently, even the Council of Ministers of the OECD adopted a declaration regarding the provision of access to research data that stems from public funding. The so-called open access movement in science and research seems to cover more and more ground. In particular, the follow-up processes of initiatives like the recent Berlin Declaration from November 2003 has started to have an influence on the academic publishing scene. Many of the central academic institutions in Europe have subscribed to this declaration that asks for open access to all research publications and opts for open archives to store and make them available. Practical measures to put this into practice are well under way. On the one hand, the worldwide database on scholarly open access journals[4] is growing by the minute. On the other hand, the big commercial publishing houses are revising their copyright policies to allow for self-archiving; ever more authors are allowed to post their manuscripts to public servers, in either a pre- or even post-publication format.[5] To sum up, the movement is already making a difference, but to make academic publishing universally open, there is still a long way to go.

Important additions to such a perspective are, for sure, further institutionalist concerns. Issues of path-dependency and sunk costs should not be underestimated. Commercial publishers as well-established actors of the present should not simply be expected to vanish, and they certainly will try to develop strategies with a view to avoiding marginalization. For instance, document delivery services, on-demand publishing and various archives are already well under way. The "battle" between the incumbents and the new actors has only just begun. In addition, we need to take into account that even the various actors within academia have different interests. Corporate research universities play a different tune to under-funded state universities, so institutional competition could be a significant impediment. Also, disciplinary differences, in particular the varying proximity of the research to the market in general, are an important factor. Finally, reappropriation requires "deep-seated cultural adjustments within the academy" (Atkinson, 2000, p. 64). However, there are already new institutions and new ways of going beyond for-profit publishing; hence stability seems challenged and new pathways on offer.

Legal Perspectives

The economic and the legal side of e-publishing are strongly interwoven. The law, in particular intellectual property (IP) law and contractual law, has been the main instrument to structure this market. While the technological setup—the freely accessible internet—seemed at first a menace to this commodification process, technology has proven to have the power to turn academic publishing into a perfected market where access to all items is strongly controlled and hence perfectly exploitable. The copyright privileges assigned to the academic world (in particular for access, copying and quoting) may be overruled by contractual restrictions, which are enforced by technological barriers. Furthermore, the worldwide trend to place universities and research institutions "in the market," that is both allowing and forcing them to become financially more independent from state resources, represents another danger: Many of the above mentioned privileges can only be enjoyed by noncommercial enterprises. To cite but one example, the U.S. fair use rules do not apply to commercial research enterprises (a similar rule is enshrined in the copyright law Directive of the EU from 2001). The more commercially oriented research institutions (universities) become, the more difficult it will be to rely on these privileges.

It is, however, doubtful whether, in the academic world, the challenges of the digital media can be adequately solved with IP regimes and licensing. Many argue that the scholarly communication system needs free access and no restriction whatsoever to the material published by other academics (see already above). Academic publications are, in most cases, no source of income for academics and the value added by publishing houses through layout, typesetting, technical markup and managing the publishing process should have a price to be paid out of university and research budgets. Furthermore, protection against fraud and plagiarism is usually provided by standard academic fraud procedures, not via copyright infringement cases. Hence, it seems conceivable in academic communication that copyright may play a much lesser role than hitherto argued from the part of the publishers. In other words, one may understand publishing academic works not as a business, but as a service from the academic community to the academic community. It does not need to be a free service, but the future system could avoid the obstruction of scholarly communication which looms if it is no longer allowed to distribute electronic (or, for that matter, print) publications freely. The new publishing system can only take place in a sort of "copyright-free zone." While IP laws would still be in force, they would not be enforced: academic authors would not care about them—and perhaps use the Open Publication Licence (OPL). What is at stake here is the choice between two potential worlds of cyberscience: the one would be ruled by statutes (copyright), contracts (licensing) and techniques (digital rights management—DRM) developed for the commercial sector; the other would be community-oriented, de-commodified and guaranteeing free access to and use of knowledge. While the first option will most likely lead to a cyberscience world in which the technological potentials are not fully exploited, the second option gives cyberspace a fair chance to develop into an adequate and improved environment for science and research. The new and much cheaper e-publishing opportunities may be an incentive and an opportunity at the same time. It would, however, need almost a revolutionary approach in many respects and by many participants. Although academia has much to gain from a financial and a practical point of view, whether this transition will take place is dependent on many further factors. In particular, whether it will indeed be possible to leave traditional paths, to overcome institutional inertia and to bypass the established incumbents has to be left open here.

Collaborative Culture and Publishing

A fifth cornerstone of the future scenario is the observation that academic work and, consequently, publishing is becoming ever more collaborative, which in turn is supported by ICT. A number of studies show that collaboration has been increasing over the last decades. For instance, scientometric data document the increase in multi-authored papers, in particular in the natural sciences (e.g., Price, 1986, 1963; Thagard, 1997). It was found that the number of international collaborative papers approximately doubled while at the same time, there was a nine-fold increase in the number of publications by large international collaborations (Walsh & Maloney, 2002, p. 3). Furthermore, the percentage of papers published with authors from more than one country significantly increased (Walsh & Roselle, 1999, p. 54). Scientific work is thus increasingly geographically distributed.

For sure, multi-authorship and the increase in distant collaboration is not unilaterally caused by CMC, but the latter contributes to and favors the former to a large extent. Most present day research obviously "needs" collaboration. There are a number of other reasons which promote the recent increase in transnational cooperations, among them funding policies, growing mobility, the increase of the overall number of researchers and of their specialization and, last but not least, content-related reasons. Furthermore, collaboration among scholars and scientists is not universal. There are considerable differences between the different fields (see already Becher, 1989; Nentwich, 2003, pp. 168 ff.). However, given these multiple pressures, even traditionally less collaborative fields increasingly tend to seek collaboration.

Collaboration is not only increasing, but collaborative patterns themselves are changing. Walsh and Roselle (1999) claim that the prior empirical work on the effects of the internet on science would suggest that scientific work is changing in profound ways. According to these authors, "the most significant change may be the transformation of collaboration patterns" (1999, 71). Among other aspects, like collaboratories (virtual places of collaboration, see Olson, Bos, & Zimmermann, 2006), we should note that new forms of collaboration emerged: Collaboration in the age of cyberscience may take the form of cooperative activities to build shared data or knowledge bases. In some fields, academics already contribute and have access to common databases, often managed by international networks (e.g., the HUGO—human genome project[6]). Increasingly, filling and structuring e-archives and databases has also become the content of whole research projects (e.g., the *Codices Electronici Ecclesiae Coloniensis* project[7]). Even more advanced would be what I mentioned above in the section on the future of scholarly publishing, namely "hyperbases" or "knowledge bases." These are potentially huge online databases filled by many in highly structured, worldwide collaboration.

A Scenario

Summing up the above considerations, we conclude that scholarly publishing is changing dramatically and will be much different from what it looks like today. It is likely that most academic writing will be published primarily in digital form. Only very few types of scholarly publications will still be printed. Academic writing will assume new shapes, in particular multimedia enhancements and hypertextual elements. It is not unlikely that the system of quality control will not only become faster, but will also benefit from the digital opportunities of both *ex-ante* and *ex-post* control. Online quality control may become much more communicative and dialogical. The trend towards more open access to academic publications, either by an author-side financed system or by institutional support and open-content licensing schemes, is likely to continue. I expect that the body of (published) knowledge will soon not only be universally accessible for free, but also networked, that is interlinked in a huge net of corresponding two-way links. Furthermore, a growing proportion of academic publications will be written by two or more, not co-located authors. Scholars and scientists will collaborate from a distance to produce new forms of knowledge representations, such as sophisticated databases.

It may well be that, while some fields will only slowly and gradually adopt the new publishing opportunities, in other fields the more revolutionary concepts may take shape in the not-so-distant future. In particular, a unified publication archive into which everything to-be-published will already go as an un-refereed preprint, seems attractive (most of physics has already implemented it, mathematics and the cognitive sciences are following). If combined with quality labelling, and if other ratings, as well as comments, were linked to each article in the database, this could turn into a user-friendly publishing system of the future, which would allow for pre-selection of quality levels. In addition, academic literature would be more embedded in the scholarly communication process.

The Need for a New Cyberinfrastructure for Academic Publishing

Whether or not one agrees with my assessment that the above scenario is likely and desirable, there can be no doubt that academia is already quite far on the path leading to this scenario. Therefore, science policy (in a wide sense, including all levels of actors, from the state to the academic institutions and associations) has to acknowledge the need to (co-)shape the scholarly e-publishing cyberinfrastructure of the future. This chapter will conclude with a few practical recommendations for this design process. As we shall see, what is needed is hardly technology, but organization, management procedures and legal as well as economic knowledge at the interface of technology and the social environment— a result that comes as no surprise in a STS study.

To begin with, by challenges for the cyber- (publishing-) infrastructure I do not understand the *hardware*. Storage capacity and network connections of academia may be not enough for many specialized applications placed under the label "e-science" (elsewhere in this volume), but it is certainly developed enough to fulfil the needs of scholarly publishing.[8] There are, however, a number of further challenges to be met.

The digital publishing environment is not yet considered trustworthy, for a variety of reasons. Besides fears of technological breakdown, legal uncertainties and quality control are the most important factors. Therefore, one significant element of the future publishing infrastructure is a stable, worldwide harmonized *legal framework* that considers academic publications in digital format a vital step of the academic communication chain. This means that legislation needs to recognize that barriers to the free flow of research communications have to be dismantled (OECD, 1998, p. 225; OECD Council, 2004). Protection against theft of intellectual achievements need not be achieved by the law and by commercial considerations, but based on the principle of *"do ut des,"* meaning the free flow of information, at least within the academic community (e.g., Bessen & Maskin, 1997; Okerson, 1997).

The other element of the emerging infrastructure relates to *quality control*. The simple transfer of the established system for securing quality of published academic work is possible (and already largely put into practice), yet not necessarily satisfactory. As we have seen, the fact that future publications will be in electronic format enables us to envisage more sophisticated, more communicative and more responsive quality control systems. The standards for these enhanced quality control mechanisms have to be established and the organizational setting to ensure these standards put in place. This is a key task for scholarly associations worldwide. It is no simple task though. Some of the promising *ex-post* quality-control tools are highly dependent on access control in order to avoid fraud. This may mean that closed groups of academics would have to be established (Atkinson, 1996, speaks of the "control-zone") in order to be sure that none is counted twice (when rating) or has the necessary credits (when commenting). Use-tracking in particular, may involve privacy problems that need to be addressed (Nentwich, 2003, 379 ff.). Also, advanced citation analysis that distinguishes between various types of citations needs standards and implementing procedures to work out properly.

A rather costly part of the future cyber-publishing infrastructure will be the digitisation of the older print-only literature. This is no luxury. An academic community that slowly moves towards an exclusively electronic publishing environment needs access to *all* academically relevant publications for direct online citation. Otherwise, the older publications will soon disappear from the radar of (not only younger) academics. In other words, although we observe a slow move towards digital publishing, it is likely that there will be a point of no return. Once passed, we will need to digitize also the entire legacy.[9]

What yet has to be established is the essential *archiving infrastructure*—a task that mainly the central libraries are about to take over, but which needs support from the academic community (UNESCO, 2003). The infrastructure in demand is more than servers with huge storage capacities. It is, first and foremost, organization of sustainable procedures, standardization of formats and decisions about selection and version control. Field-wide open archives or repositories that store practically all published material could be a good starting point for long-term archiving endeavors.

Speaking of *software*, a main part of the infrastructure needs to be handy tools to write, coauthor and organize the new publishing formats. Although there exist prototypes of hypertext editors and professional multimedia tools, etc. there is not yet software out there that is comparable to standard text editors when it comes to user friendliness. As long as the software which academic authors use is not able to support in an easy way the preparation of metadata, including keywords, to link to other publications in a structured and harmonized way, etc., these promising paths will not be taken.

Conclusion

Cyberscience is in the making, as is the cyberinfrastructure for scholarly publishing of the future. STS certainly has a role in describing and analyzing this development in a retrospective manner. In addition, STS researchers interested in future developments, among them those belonging to the community of *constructive* technology assessors in a wide sense, add a further dimension. As long as a new technology, together with its organizational environment, is not yet fully established, we can not only look at potential consequences for the area of application (here: for academia), but we can also contribute to the very "construction" process. This chapter went along this route by drawing practical conclusions in the form of recommendations for those involved at all levels. Constructive technology assessment (cTA, see, e.g., Rip, Misa, & Schot, 1995), however, would always involve those directly concerned by the new technology. This means in the case at hand that representatives of scholarly associations, academic librarians, publishers, university administrators and, above all, scholars, scientists and researchers need to be included. Although the actual shaping of the technology takes place on the ground—by researchers using it in a specific way and giving feedback to those developing it further—decisions concerning the framework and the infrastructure are mostly taken at levels remote from the users. So far, we have not seen intensive debates, organized feedback-loops and co-construction in most academic fields. STS in terms of cTA is contributing by mediating and evaluating this fundamental process of change.

References

Atkinson, R. (1996). Library functions, scholarly communication, and the foundation of the digital library: Laying claim to the control zone. *The Library Quarterly, 66*(3), 239-265.

Atkinson, R. (2000). A rationale for the redesign of scholarly information exchange. *Library Resources and Technical Services, 44*(2), 59-69.

Becher, T. (1989). *Academic tribes and territories*. Milton Keynes: Open University Press.

Bessen, J., & Maskin, E. (1997, 1997-01-23/25). *Intellectual property on the internet: What's wrong with conventional wisdom?* Paper presented at Internet Publishing and Beyond: Economics of Digital Information and Intellectual Property, Cambridge, MA. Retrieved July 25, 2005, from http://www.researchoninnovation.org/iippap2.pdf

Harnad, S. (1990). Scholarly skywriting and the prepublication continuum of scientific inquiry. *Psychological Science, 1*, 342-343. Retrieved July 21, 2005, from http://www.cogsci.soton.ac.uk/~harnad/Papers/Harnad/harnad90.skywriting.html

Hey, T., & Trefethen, A. E. (2002). The UK e-science core programme and the Grid. *Lecture notes in computer science* (Springer-Verlag), 2329(3). Retrieved on July 21, 2005, from http://www.ecs.soton.ac.uk/~ajgh/FGCSPaper.pdf

Keller, A. (2001). *Elektronische Zeitschriften im Wandel: Eine Delphi-Studie*. Wiesbaden: Harrassowitz.

Mogge, D. (1997). Foreword. In *7th ed. of the directory of electronic journals, newsletters and academic discussion lists*. Retrieved August 13, 2002, from http://db.arl.org/foreword.html

Nentwich, M. (1999). The European research papers archive: Quality filters in electronic publishing. *Journal of Electronic Publishing, 5*(1). Retrieved July 21, 2005, from http://www.press.umich.edu/jep/05-01/nentwich.html

Nentwich, M. (2001). (Re-)De-commodification in academic knowledge distribution? *Science Studies, 14*(2), 21-42. Retrieved July 21, 2005, from http://eiop.or.at/mn/ScSt2001.pdf

Nentwich, M. (2003). *Cyberscience: Research in the age of the internet*. Vienna: Austrian Academy of Sciences Press.

Nentwich, M. (forthcoming). Cyberscience: Modelling ICT-induced changes of the scholarly communication system. *Information, Communication & Society (iCS), 8*(4).

OECD. (1998). *The global research village: How information and communication technologies affect the science system*. In *Science, technology and industry outlook 1998 (chap. 7)*. Paris: Organisation for Economic Co-operation and Development.

OECD Council. (2004). *Declaration on access to research data from public funding*. Retrieved July 21, 2005, from http://www.oecd.org/document/ 0,2340,en_2649_34487_25998799_1_1_1_1,00.html.

Okerson, A. S. (1997). *Introduction to the 6th ed. (1996) of the directory of electronic journals, newsletters and academic discussion lists*. Retrieved July 21, 2005, from http://www.people.virginia.edu/~pm9k/libsci/ 96/intro.html

Olson, G., Bos, N., & Zimmermann, A. (Forthcoming). *Science of Collaboratories*.

Price, D. J. d. S. (1986). *Little science, big science—and beyond* (Rev.ed.). New York: Columbia University Press.

Rip, A., Misa, T. J., & Schot, J. (Eds.). (1995). *Managing technology in society—The approach of constructive technology assessment*. London: Cassell Publishers Ltd.

Rogers, E. M. (1995). *Diffusion of innovations* (4th ed.). New York/London: The Free Press.

Thagard, P. (1997). Collaborative knowledge. *Noûs, 31*(2), 242-261. Retrieved July 25, 2005, from http://cogsci.uwaterloo.ca/Articles/Pages/Collab.html

UNESCO. (2003). *Charter on the Preservation of Digital Heritage*. Retrieved July 21, 2005, from http://unesdoc.unesco.org/images/0013/001331/ 133171e.pdf#page=80

Walsh, J. P., & Maloney, N. G. (2002). Computer network use, collaboration structures and productivity. In P. Hinds & S. Kiesler (Eds.), *Distributed work* (pages from manuscript). Cambridge, MA: MIT Press.

Walsh, J. P., & Roselle, A. (1999). Computer networks and the virtual college. *Science Technology Industry Review (OECD), 24*, 49-78.

Endnotes

[1]   I define "cyberscience" as all scholarly and scientific research activities in the virtual space generated by the networked computers and by advanced information and communication technologies, in general (ibid., 22).

[2]   See Mogge (1997) and the Directory of Scholarly Electronic Journals and Academic Discussion Lists (DSEJ) 2000 at <http://dsej.arl.org/>; among them are at least 1,440 that provide open access according to the Directory of Open Access Journals (DOAJ) at < http://www.doaj.org> in February 2005.

[3]   Like TRANS < http://www.inst.at/trans/>

[4]   <http://www.doaj.org>

[5]   This is well documented in the ROMEO database < http://www.sherpa.ac.uk/romeo.php>

[6]   <http://gdbwww.gdb.org>

[7]   <http://www.ceec.uni-koeln.de>

[8]   This is probably even true for most academic institutions in less developed regions of the world.

[9]   Preferably, digitisation should also include transforming into digital text, not only image scanning with metadata descriptions.

Chapter X

On Web Structure and Digital Knowledge Bases:
Online and Offline Connections in Science

Alexandre Caldas
Oxford Internet Institute, University of Oxford, UK

Abstract

This chapter analyses the web as a complex system of interactions bridging online and offline communities in open science in order to discuss the transformation of communication and practices within scientific communities. It addresses the problem of mapping the structural linkages of research networks on the internet for purposes of identifying digital knowledge bases on electronic networks. Traditional (nonelectronic) research networks are likely to have a digital representation (web presence), whose boundaries and characteristics require a closer investigation. It is of special concern here to identify particular subsets of these digital networks whose properties are related to non-digital collaboration structures. Empirical evidence for electronic connectivity on the internet is discussed from a European Language and Speech Network, constituted by 141 research groups—the ELSnet network. We explore the possibility of identifying particularly intensive "Digital Knowledge Bases" on these electronic networks.

Introduction

Traditional (nonelectronic) research networks and scientific communities more generally are likely to have a digital representation (web presence), whose boundaries and characteristics require a closer investigation. Extensive research has now been conducted on collaboratories (see Vann and Bowker in this volume for an introduction) which provide a good indication of the potentiality of ICT and e-infrastructures for research collaboration. In-depth studies have also been conducted on the use of the internet within the European research arena (see, for example, Barjak in this volume). It is our special concern here to identify particular subsets of these digital networks whose properties are related to non-digital and offline collaboration structures.

In the first place, we will be interested in testing the hypothesis that the patterns of connectivity of these electronic networks are structurally similar to the collaboration patterns identified previously by bibliometric and network analyses. Those analyses were based on research project collaboration structures that have evolved after a decade of European Commission funding of this research field and the interpersonal collaboration structures as revealed by the survey of researchers in ELSnet (a European Language and Speech Network, constituted by 141 research groups). We find that those electronic connectivity patterns allow one to identify a restricted set of "best-connected" research institutions, and the linkages to important external entities and electronic resources.

Secondly, when analyzing more deeply the inner structure of these "web communities," we can identify different degrees of centrality and prestige even among the "core" of electronically very-well-connected research institutions. At this point, we need to obtain a better understanding of how many other institutions are directly linking to each institution as well as "who is" directly connected to each of the 141 ELSnet research groups. As a complementary measure of the centrality of each institution in the whole network, we also mapped the electronic "ego-networks" (direct and indirect links starting in a specific institution and spreading into the wider internet space) for some central research institutions. Again, these centrality results corroborate the collaboration patterns of the actual nonelectronic network.

Thirdly, having characterized the connectivity pattern of these electronic networks, and the heterogeneity in centrality of different institutions within these networks, we explore the possibility of identifying particularly intensive "digital knowledge bases" on electronic networks. By mapping ego-networks of a number of better connected and highly central institutions we might be able to identify "overlapping regions" of very intensive knowledge resources topically related to this field of research. A conceptual model is presented in the final section.

The analysis is organized in the following way. In the first section we situate the topic of the "link structure" of the internet and the discovery of knowledge bases, giving an overview of the literature in this area. Sections two to four describe, respectively, the research question and hypotheses, the offline collaboration patterns and the methods used for conducting the overall analysis. The following three sections correspond to the detailed analysis and discussion of the three topics outlined above. And finally we conclude the chapter by summarizing the most significant results and limitations of this investigation and open avenues for further research.

The Link Structure of the Internet and Digital Knowledge Bases

In this investigation we will be using techniques of link-structure analysis for determining the electronic connectivity and centrality of research institutions in the overall network, as well as devising new models for discovering knowledge resources on these sparse networks. This is closely linked with search-engine research as well as recent studies exploring the topology of the internet to identify web communities. We will examine these two topics before moving on to a more sociological exploration of this problem.

Search-engine research is mainly concerned with providing a solution to two related functionalities: *precision/relevance* and *recall*. The results of a search on the internet should be as precise as possible in order to be effective. From the enormous, and perhaps impossible to determine, collection of information available, it should provide in a few seconds more precise results on the topics being searched. This goal is sometimes at odds with the number of items of information that it can give as an output to the user (the recall capacity of the search engine). We would expect that the *more* results returned, the better. In other words, search engine technology should simultaneously provide more items and very precise items of information as an outcome of any search.

Considering the size and dynamic nature of the web, this is a very complex technological undertaking. Most search engines index documents in a database recovered from systematic crawls of the web. The crawls explore the link structure and hypermedia nature of the internet in order to link from one document to related documents. Complementary content analysis provides a classification of the retrieved documents. However refined search-engine technology has come to be, estimates suggest that 12 search-engines taken together index less than 50% of the whole internet system (Lawrence & Giles, 1999).

As noted by Aquino and Mitchell (2001), search-engine technology has used three different types of algorithms to build these indexed databases: the Naïve Bayes model, which focuses on topic-word frequencies; "maximum-entropy" algorithms, which focus on word combinations and how frequently they are associated; and perhaps the most promising approach, the "co-training" model, which studies the information on a web page, as well as the linked pages, building an association of correlations. In fact, these strategies of combining the content and resources of the internet with the link structure of those resources are better suited for "entity extraction," or the ability to build databases from collections of specific entities.

This brings us closer to the notion of "web communities" and the self-organization of the internet (Kumar et al., 2000). Considering a web community as a set of web pages that link (in either direction) to more web pages inside the community than to pages outside of the community (Flake et al., 2000), it is relevant to identify such subsets of the large electronic network. Several studies have empirically identified such communities based on a combination of several algorithms (e.g., Kumar et al., 1999). One of the most commonly used is the HITS algorithm (Kleinberg, 1998) which explores the link structure of the internet, starting from a set of seed URLs and determines the *Hubs* and *Authorities* resources on internet web space. *Hubs* are internet resources that link to many authoritative pages in the topic, while *Authorities* are internet resources that are linked by many Hubs. There is a self-recurring mechanism in the identification of Hubs and Authorities. This problem has been overcome by refinements to the initial algorithm. The method is based on partitioning of the initial graph into well-defined components. The determination of Authoritative pages on a particular topic is quite useful when ranking results of search queries. The PageRank algorithm, implemented in the search-engine Google, uses a very refined version of the initial HITS algorithm (Brin & Page, 1998; Huang, 2000). Here we are going to use the Google repository for determining our collection of *Related* pages to our initial research network.

From a technical point of view, theoretical and empirical research have demonstrated the effectiveness of using the link structure of the Network to identify subsets of the larger network that can be categorized as "web communities" (Cooley et al., 1997; Kosala & Blockeel, 2000). However, the analysis of the inner structure of these "web communities" is still a challenge, and probably the solution is not particularly technical in nature but more sociological. In fact, we should try to understand the organizations of these electronic communities by comparison with other structural characteristics, such as the similarity of the information under exchange within these communities, or the offline collaboration structure of the institutions participating in these communities. A more general background on network analysis is provided by Wasserman and Faust (1994) and Moreno (1932).

Sociological explanations of online virtual communities have particularly focused on user studies of the behavior of those communities when participating in electronic environments (Wellman, 1996; Garton et al., 1999; and Koku et al., 2000). Much less research has been done on the automatic examination of electronic collaboration structures, as given by internet resources and archives, at a level higher than the individual, and not focused on particular events (such as electronic conferences or newsgroup participation), but unrestricted from the point of view of having a particular causal factor for their occurrence.

Studies seeking to explain the sociological factors underlying the electronic connectivity of a set of institutions are rare. In fact, it is the particular focus of this research to start the analysis with a well-bounded and restricted set of research institutions (that are relevant for participating in the same research network). This focused search is likely to impose an *a priori* restriction on the total electronic webspace, which might yield very good results in terms of topical focus and discovery of knowledge resources. In addition, we explore the idea that the electronic webspace of the community should in a particular way be related to the sociological structure of the community. Connectivity patterns and the centrality of some institutions within the electronic network should also reflect this. These are all research questions not covered by the existing literature and where there are important theoretical hypotheses to be investigated.

Hypotheses to be Investigated

A crucial research question is whether the electronic patterns of connectivity of a certain set of research institutions reflect the structures of research collaboration (Katz & Martin, 1997) of the offline (nonelectronic) research network. If so, what kind of structure is revealed by the electronic network? Can we identify the best-connected institutions and what characteristics they have?

Related to this set of questions is the possibility of discovering very intensive knowledge resources on electronic networks by mapping the topology of densely-connected and central institutions in internet space—we will refer to these as "digital knowledge bases."

The analysis is conducted with a view to testing the following theoretical hypotheses:

a. The pattern of electronic connectivity for the set of research institutions is structurally similar to the already analyzed offline collaboration structures within the ELSnet network.

    This should allow one to identify a core of better-connected institutions, also in electronic space, as well as to distinguish within this group certain properties explaining this connectivity behavior. Moreover, the network mapping of the electronic relationships should allow one to identify institutions that collaborate more intensively with each other in traditional settings.

b. Even among the better connected "core" group of institutions, we are able to identify different degrees of centrality and prestige, as measured by the number and type of institutions connected to them (*indegree)* and their activity in electronic space—the number and type of links originating from them (*outdegree*).

    This allows one to identify some heterogeneity within the network characterizing the inner structure of the electronic map. This leads us to a final hypothesis related to the potential identification of knowledge bases in electronic networks.

c. The mapping of the ego-centered networks (the links originating in a certain institution and extending into a specified depth) of the most central and best-connected institutions allows the discovery of digital knowledge bases.

Assuming that hypotheses a and b are validated, that is, the electronic connectivity resembles the collaboration pattern of the institutions involved, and that there is the possibility to distinguish the most central and more active institutions, then we might explore the idea that the detailed examination of their electronic links will allow one to find important and topically related knowledge resources on the internet. As this final hypothesis is exploratory, we discuss here a conceptual model for the identification of these "Digital Knowledge Bases."

Offline Collaboration in Speech and Language Research

More traditional and offline collaboration patterns in speech and language research, as given by co-authorship in publications as well as by collaboration in European research projects during a period of two decades, have been previously analyzed (Caldas, 2004). An analysis of collaboration in funded research projects by the European Commission provided the basis for identification of stable groups of collaborators in speech and language. These programs of research include projects of the Framework Programmes 3 and before that (in a period prior to 1994) such as the Language Engineering Programme (1994-1998) and the Human Language Technologies Programme (1998-2002). A second group, with a more diversified set of research objectives and not directly oriented to speech and language, includes the research projects conducted under the ESPRIT programme (1994-1998), the Multilingual Information Society Programme (1996-1999), the INCO International Co-operation Programme (1994-1998) and the E-Content Programme (2001). These research programmes were oriented towards the production and dissemination of multimedia content, translation services and the application of multilingual services in the information society. The content nature of the research programmes was found to be clearly reflected in the structure of the corresponding networks. The difference is significant in terms of the participating institutions, the connectivity of the networks and the collaboration groups formed within these networks. Moreover, a closer analysis of the persistence of collaboration groups within these networks, and particularly an analysis focusing on the specialized networks dedicated to speech and language research, appears to confirm the continued presence of certain institutions, as well as a long-term collaboration pattern over time.

Table 1 summarizes those research institutions that participated actively in all these research programmes, and are persistent and stable members of these offline collaboration networks. In later sections of this chapter we will find out that a significant proportion of these institutions are also well-connected in the online environments of the web.

A second trace of collaboration in the offline world has been taken from a long-term (20 years) bibliometric study of co-authorship in speech and language. The bibliometric analysis is based upon collection and networks analysis of articles published in speech and language (from ISI Science Citation Index web database, 1981-2000). This research used the concept of "collaboration group" for a group of research entities, whose researchers (as identified in the publication records) had over time coauthored papers together and formed

Table 1. Group of research institutions with long-term participation in speech and language networks

Long-term Active Institutions in Speech and Language Networks
CLI—Computational Linguistics Institute—Italy
CPK—University of Aalborg—Denmark
CST—Centre for Speech Technology—Denmark
DFKI—German Research Centre for Artificial Intelligence—Germany
ILSP—Institute for Language and Speech Processing—Greece
INESC—Instituto Engenharia de Sistemas e Computadores—Portugal
LIMSI—Laboratoire d'Informatique Mécanique et les Sciences de l'Ingénieur—France
KUN—Centre for Language and Speech—Netherlands
KTH—Royal Institute of Technology—Sweden
KULeuven—Speech Group of Katholieke Universiteit Leuven—Belgium
IGFAI—Institute for Artificial Intelligence—Germany
UPC—Polythecnique University of [C]atalunia—Spain
IMS—University of Stuttgart—Germany
University of Sheffield—United Kingdom
University of Utrecht—Netherlands
University of Edinburgh—United Kingdom
Rank Xerox—France

"cliques" of coauthors. The reliable identification of research groups by coauthor analysis required the combination of several complementary strategies (for bibliometrics background, see, e.g., Van Raan, 1998). This situation was augmented when identifying groups in interdisciplinary fields of research (Bordóns, et al., 1995) like this one. Among the techniques, it was necessary to use a combination of: (1) an Extended Broad Search method which required *keyword* search, *key-journals* search and *key authors* search, for the delimitation of the bibliometric data set; (2) the separate analysis of bibliographic-coupled documents with important works in the fields of speech and language research; and (3) the long-term analysis of the evolution of the co-authorship networks.

The combination of these methods allowed the identification of the following groups within the European Speech and Language area. Some of those "collaboration groups" are listed in Table 2, indicating some highly visible international groups. The results reinforce those discussed above for collaborating research groups in European research projects, since a good proportion of these research groups identified by bibliometric analysis was found to be significantly well-connected in electronic networks and the web.

There is no relevance in the relative ordering of the above institutions. The important point to emerge from the offline empirical evidence is that all of them are among a "core" group of institutions collaborating more often within these research networks over a significant period of time (1981-2002). The next sections will examine whether the electronic networking of research reproduces some of these patterns.

Table 2. Selected research groups identified by co-authorship analysis

Research Groups identified by co-authorship analysis in Speech and Language (1981-2000)

Speech Research Groups in Europe
Univ Amsterdam, Inst Phonet Sci, IFOTT, NL-1016 CG Amsterdam, Netherlands
Departamento de Sistemas Informáticos y Computación, Universidad Politécnica, Valencia, Spain
Univ Granada, Fac Ciencias, Dept Elect & Tecnol Comp, E-18071,Granada, Spain
Univ Nijmegen, Dept Language & Speech, A2RT,POB 9103, NL-6500, HD Nijmegen, Netherlands
Lab Linguist Formelle, Case 7003, 2 Pl Jussieu, F-75251 Paris, France
Univ Paris 07, CNRS, F-75251 Paris, France
Univ Cambridge, Dept Engn, Cambridge CB2 1PZ, England
Philips Res Labs, POB 1980, D-52021 Aachen, Germany
Siemens Brussels, Adv Proc Control Grp, Huizingen, Belgium
CNRS, LIMSI, Spoken Language Proc Grp, BP 133, F-91403 Orsay, France
Univ Bonn, Comp Sci Dept 3, D-5300 Bonn, Germany
Georgian Techn Univ, Dept Digital Commun Theory, GE-380075 Tbilisi, Rep of Georgia
Univ Surrey, Ctr Commun Syst Res, Guildford GU2 7XH, Surrey, England
Language Research Groups in Europe
Max Planck Inst Cognit Neurosci, D-04103 Leipzig, Germany
EHESS, CNRS, Lab Sci Cognit & Psycholinguist, F-75651 Paris 13, France
Univ Karlsruhe, AIFB, Kaiserstr 12, D-76128 Karlsruhe, Germany
Univ Erlangen Nurnberg, Chair Pattern Recognit, Martensstr 3, D-91058 Erlangen, Germany
Bavarian Res Ctr Knowledge Based Syst FORWISS, D-91058 Erlangen, Germany
Univ Freiburg, Computat Linguist Lab, D-79085 Freiburg, Germany
MRC, Cognit & Brain Sci Unit, Cambridge CB2 2EF, England
Univ Nijmegen, Dept Exptl Psychol, 6500 HE Nijmegen, Netherlands
Univ Maastricht, MATRIKS, Dept Comp Sci, Maastricht, Netherlands
DFKI, D-66123 Saarbrucken, Germany
Leiden Univ, Dept Comp Sci, POB 9512, NL-2300 RA Leiden, Netherlands

Methods for Mapping Research Networks on the Internet

The empirical analysis will be focused on the examination of the electronic connectivity of the member institutions of the ELSnet network—The European Speech and Language Research Network. This network comprises a set of 141 research groups/institutions across 27 countries all over Europe. Most of these institutions were the ones participating in the Research Programmes and Projects funded by the European Commission over the period 1990-2002, whose structural detailed analysis has been conducted by the author (see previous sections). These collaboration structures, identified previously by network analysis, will be used as a proxy for the collaboration structure within the whole set of institutions of ELSnet. Moreover, the ELSnet network researchers were the population used for a survey on collaboration and use of information and communication technologies (Caldas, 2004). The interpersonal collaboration structures, as given by the institutions with whom the researchers collaborate more, will also be used as a proxy for the collaboration structures within the network.

The analysis of the electronic connectivity of the ELSnet institutions is based on data collected from the search engine Google (reachable at http://www.google.com). This search engine technology was chosen for three reasons: (1) This search engine repository is considered one of the largest indexed databases (Search Showdown, 2002); (2) the technology and algorithms underlying the search process are covered in the literature and particularly suitable for the analysis of web communities; and (3) two specific functionalities of this search engine (advanced search of the pages that link to a specific page—*Link:* feature; and advanced search of a set of approximately 30 pages that are strongly related to a specific page—*Relate:* feature) were only available from this search engine, at the time of analysis.

The methodology (for a comparable analysis see, e.g., Flake et al., 2000, 2002), based in a combination of webmetric techniques and network analysis, consisted of the following procedures:

- To collect and analyze for each research institution all the other institutions that are strongly related with them (as given by the search-engine technique described above);

- To carry out a detailed network analysis of the connectivity of the resultant network of interrelated research institutions (in webspace);

- To compare the electronic connectivity network with the patterns of collaboration revealed by the physical (nonelectronic) ELSnet network;

- To collect and analyze for each research institution all the institutions that are linked directly to it (as given by the link: pages feature of Google as described above);

- To characterize the centrality and prestige of the research institutions as given by their position in the electronic network as well as the total number of links pointing to them and the type of institutions linking to them;

- To collect data for the ego-centered electronic networks of a sample of very well connected and central research institutions; and

- To produce a network map of these ego-networks in order to have a visual inspection of their treelike structure and to derive the conceptual model of the final section.

The discussion of the results follows in the next three sections, dealing respectively with the patterns of electronic connectivity, the centrality and prestige of research institutions in electronic space and the conceptual model for the discovery of digital knowledge bases.

Patterns of Electronic Connectivity of Research Networks

We should outline some methodological restrictions from the beginning. From the total set of 141 research institutions participating in the ELSnet network, we were only able to analyze the patterns of electronic connectivity of 108 of these research institutions. The remaining 33 institutions were not analyzable for several reasons: (1) 13 institutions did have a URL identification, but the URL referred to the global institution and not specifically to the research centre or research group focused on speech and language research or development. This was usually the case of large R&D companies in this field, for example., Nokia research, IBM Deutschland, DaimlerChrysler AG, Philips Electronics, BT Adastral Park, but also some university faculties, e.g., University of Lisbon—Department of Informatics and Università degli Studi di Pisa; (2) 12 institutions did not have a URL, at least as given by the members list of ELSnet members. In some of these cases, they probably do have a URL but it is too generic to represent the specific research group. (3) Eight of the institutions did not have any relationship result, as given by the Google *relation:* functionality.

For the analyzed research institutions (108), we were able to determine a list of related entities (URLs). This list of related entities is restricted to a maximum of 30 URL references. These references point to strongly-related internet resources (web pages). The measure of relatedness is given by a combination of structural relationship—sets of linkages characterizing two related institutions, as well as topic/content similarity. For the present purposes, we consider this indicator of relatedness as a measure of electronic connectivity between institutions. It should be stressed that the list of related institutions is not restricted to the members of the ELSnet network. Even if our main purpose is the identification of the connectedness between members of the network, most of the "links" of relatedness were to outside entities and URL references. In fact, some highly relevant external links to "authoritative" URL references are one important characteristic of the best-connected institutions (see more about this on the final part of this section). For an analysis of the structure of the web and its significance, see, for example, Huberman et al. (1998), Kleinberg (1998) and Kleinberg and Lawrence (2001). For models estimating the size of the web, see, for example, Albert and Barabási (1999).

From the analysis of this connectivity dataset we were able to identify a "core" set of 24 institutions with a denser relationship with other members of the ELSnet network. A combination of three important properties distinguishes this reduced set of institutions. First, as referred above, all of them reveal a more intensive pattern of electronic relationship with other members of the ELSnet research network. Secondly, they also have electronic relationships with other institutions not part of the ELSnet network, but working in the same research and development field. Thirdly, as noted above, they have electronic relationships with important resources which are external to the network but are considered hubs and authorities (in the sense as defined in the literature review), that,

Table 3. The 24 research institutions more densely connected, as given by electronic connectivity

Best Electronically Connected Institutions	
Austrian Research Institute for Artificial Intelligence (**Oefai_at**)	http://www.ai.univie.ac.at/oefai/nlu
Katholieke Universiteit Leuven—Centre for Computational Linguistics (**kuleuven_be**)	http://www.ccl.kuleuven.ac.be
Aalborg University—Institute of Electronic Systems—Center for Sprogteknologi (**Cpk_dk**)	http://www.cpk.auc.dk
Inst. National Polytechnique de Grenoble—Institut de la Communication Parlee (**Grenet_fr**)	http://www.icp.grenet.fr
LIMSI/CNRS—Human-Machine Communication Department (**Limsi_fr**)	http://ww.limsi.fr
Universite de Provence—Laboratoire Parole et Langage (**Univ-aix_fr**)	http://www.lpl.univ-aix.fr
Christian-Albrechts University, Kiel, Inst. Phonetics Digital Speech Processing (**Uni-kiel_de**)	http://www.ipds.uni-kiel.de
German Research Centre for Artificial Intelligence—Language Technology Lab (**Dfki_de**)	http://www.dfki.de/lt
Institut fur Angewandte Informationsforschung (**Iaiuni-sb_de**)	http://www.iai.uni-sb.de
Universitaet des Saarlandes—Department of Computer Linguistics (**Coliuni-sb_de**)	http://www.coli.uni-sb.de
Universitat Stuttgart-IMS—Institut fur Maschinelle Sprachverarbeitung (**Ims_de**)	http://www.ims.uni-stuttgart.de
Consiglio Nazionale delle Ricerche--Instituto di Linguistica Computazionale (**Ilccnr_it**)	http://www.ilc.pi.cnr.it
Royal Institute of Technology—Department of Speech, Music and Hearing (**Kth_se**)	http://www.speech.kth.se
University of Geneva—ISSCO/ETI (**Issco_ch**)	http://www.issco.unige.ch
University of Nijmegen—Department of Language and Speech (**LetKun_nl**)	http://lands.let.kun.nl
Utrecht University—Utrecht Institute of Linguistics OTS (**Letuu_nl**)	http://www-uilots.let.uu.nl
UMIST—Computational Linguistics (**Umist_uk**)	http://www.ccl.umist.ac.uk
University College London—Department of Phonetics and Linguistics (**Phonucl_uk**)	http://www.phon.ucl.ac.uk
University of Cambridge—Computer Laboratory (**Clcam_uk**)	http://www.cl.cam.ac.uk
University of Cambridge—Speech, Vision and Robotics Group (**Engcam_uk**)	http://svr-www.eng.cam.ac.uk
University of Edinburgh—Informatics (**Cogscied_uk**)	http://www.cogscied.ed.ac.uk
University of Sheffield—ILASH, Computer Science (**Dcsshef_uk**)	http://www.dcs.shef.ac.uk/research/ilash
University of Sussex—School of Cognitive and Computing Sciences (**Cogssusx_uk**)	http://www.cogs.susx.ac.uk/lab/nlp/index.html

internet resources pointed to by many other URL references and linking to other authoritative resources.

Table 3 shows the list of the best electronically connected institutions. A comparison of this list of institutions with the research institutions appearing in the collaboration structures identified by network analyses—structural patterns of research collaboration as given by research funding of this field by the European Commission, as well as by interpersonal collaboration maps resulting from the researchers survey and co-authorship analysis—reveals a very good similarity.

However, one characteristic is that these are research institutions from university departments or government laboratories. Thus, to a very large degree, we were able to identify from the electronic connectivity patterns, the best-connected "open science" research institutions, but not the R&D companies involved in this research area. This is probably due to the methodological constraint referred to initially, and also because of the "higher level of closeness" of these companies in publicly revealing their research information. This might represent a problem in the form of an electronic under-representation of this group in terms of any structural analysis of large-scale systems such as the internet.

Figure 1. Visual representation of electronic connectivity network (24 core members of ELSnet network)

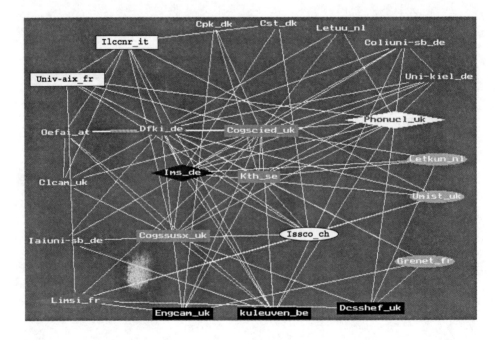

In order to map the connectivity pattern of this reduced network, we did a network analysis of the electronic connectivity. The results are represented in Figure 1.

From the above network map, it is clear that all the institutions reveal a reasonable degree of connectedness with the other members of the network. Nonetheless, it is also possible to visualize the heterogeneity within this reduced set, in terms of degree of connectedness and centrality. For example, institutions such as the Universität Stuttgart-IMS—Institut für Maschinelle Sprachverarbeitung (*Ims_de*), the German Research Center for Artificial Intelligence (DFKI)—Language Technology Lab (*Dfki_de*), the KTH (Royal Institute of Technology)—Department of Speech, Music and Hearing (*Kth_se*) and the University of Edinburgh—Informatics (*Cogscied_uk*)[1] occupy important positions within the network. They bridge the connection between subgroups interacting in the network. This higher degree of centrality is further discussed in sections that follow.

A final comment on the results for the pattern of electronic connectivity of these best-connected institutions is their linkage with external resources. A major

Table 4. Structurally significant external connections (hubs, authorities and other internet resources)

Hubs, Authorities and Internet Resources	
Hubs	
ELSnet Network	http://www.elsnet.org
ELRA/ELDA (European Language Resources Association)	www.icp.inpg.fr/ELRA/home.html
HLT (European Commission Human Language Technologies)	www.hltcentral.org/page-56.shtml (EUROMAP)
ISCA (International Speech Communication Association)	www.isca-speech.org/
Linguistic Data Consortium	www.ldc.upenn.edu/
ETSI (European Telecommunication Standards Institute)	www.etsi.fr/
International Phonetics Association	www2.arts.gla.ac.uk/IPA/ipa.html
American Speech Language and Hearing Association	www.asha.org/
Association for Computation Linguistics	www.aclweb.org/
Authorities	
Stanford Linguistics Research	www-linguistics.stanford.edu/
SRI Natural Language Program Research	www.ai.sri.com/natural-language/
Ctr.Language & Speech Processing at John Hopkins University	www.clsp.jhu.edu/index.shtml
Perceptual Science Lab	mambo.ucsc.edu/
Speech at Carnegie Mellon University	www.speech.cs.cmu.edu/speech/
Microsoft Natural Language Processing Home	research.microsoft.com/nlp/
University of Rochester Department of Linguistics	www.ling.rochester.edu/
MIT Artificial Intelligence Lab	www.ai.mit.edu/
Speech Recognition Group at Rutgers University	www.caip.rutgers.edu/ARPA-SLT/
Resources	
PRINT SERVER Computational Linguistics	xxx.lanl.gov/cmp-lg/
VIRTUAL Library of Linguistics	www.emich.edu/~linguist/www-vl.html
Comp Speech Newsgroup Website	www.speech.cs.cmu.edu/comp.speech/
Subgroup of the Association for Computation Linguistics	www.sigdial.org/
Natural Language Translation Specialist Group	www.bcs.org.uk/siggroup/nalatran/nalasupp.htm
The Linguist List - Mailling List	www.emich.edu/~linguist/
Stanford Grammar Resources	hpsg.stanford.edu/

proportion (in all but two) of the 24 best-connected institutions was related to external internet resources. The striking result is that even within the reduced set, the more densely connected institutions, such as the ones referred above, pointed to the same "authoritative" institutions giving credibility to the *reinforcing mechanism* statement that the more authoritative resources tend to be reinforced by linkages with other authoritative resources. The better connected one is, the more one is likely to be structurally related with other very well-connected internet resources. In this sense, a form of electronic "Matthew Effect" is likely to occur within these electronic networks. Table 4 summarizes the most relevant "external" linkages.

It is significant that we were able not only to identify the best-connected institutions by analysing the electronic connectivity of the initial research network, but also that these best-connected entities are electronically connected to important topic-related external resources on the larger internet network. This is particularly significant when discussing the potential discovery of digital knowledge bases by mapping the electronic connectivity ego-maps of the most relevant institutions. Moreover, it is a good indication of heterogeneity in the kind of knowledge resources we are able to identify when starting the analysis with a close-knit, bounded research network, and then expanding the electronic network to include "external" highly relevant internet resources.

In order to gain a better understanding of the inner structure of these electronic networks, we will discuss in the next section centrality and prestige measures of the institutions under analysis. We will be interested in differentiating the institutions in terms of their activity and prestige, even among the more reduced core set of "best-connected" institutions.

Centrality and Prestige within Electronic Networks

Measuring the Prestige of Research Institutions on Internet Web Space

A common indicator of the "prestige" of a certain entity within a network is given by the total number of links that entity/institution receives from other institutions within the network. Table 5 shows the results for the indegree centrality of the already selected best-connected institutions within the ELSnet network. It should be noted that these include incoming links from other institutions outside the network, as well as "inside" links from within the same institution (an

Table 5. Indegree-centrality of ELSnet best-connected institutions

Best Electronically Connected ELSnet Institutions - Indegree Links	
Research Institutions	*Number of Direct InLinks*
Austrian Research Institute for Artificial Intelligence (OFAI)	80
Katholieke Universiteit Leuven - Centre for Computational Linguistics (CCL)	258
Aalborg University - Institute of Electronic Systems	486
Center for Sprogteknologi	126
Inst. National Polytechnique de Grenoble - Institut de la Communication Parlée	344
LIMSI/CNRS - Human-Machine Communication Department	1660
Université de Provence - Laboratoire Parole et Langage	866
Christian-Albrechts University, Kiel - Institute of Phonetics and Digital Speech Processing	70
German Research Center for Artificial Intelligence (DFKI) - Language Technology Lab	498
Institut für Angewandte Informationsforschung	144
Universitaet des Saarlandes Department of Computer Linguistics	1030
Universität Stuttgart-IMS - Institut für Maschinelle Sprachverarbeitung	1090
Consiglio Nazionale delle Ricerche - Istituto di Linguistica Computazionale	306
KTH (Royal Institute of Technology) - Department of Speech, Music and Hearing (TMH)	1050
University of Geneva - ISSCO/ETI	454
University of Nijmegen - Department of Language and Speech	86
Utrecht University - Utrecht Institute of Linguistics OTS	534
UMIST - Computational Linguistics (old URL)	234
University College London - Department of Phonetics and Linguistics	974
University of Cambridge - Computer Laboratory	2180
University of Cambridge - Speech, Vision and Robotics Group	1060
University of Edinburgh - Informatics	542
University of Sheffield - ILASH, Computer Science	334
University of Sussex - School of Cognitive and Computing Sciences	160

indicator of "inbreeding", or self-citation) and as such it should be taken as merely a proxy for the importance of the institutions in electronic space in general, and not as a measure of the network in particular.

The results corroborate the "centrality" and "prestige" of the best-connected institutions, and this applies to the ones with higher degree of connectivity as given by the analysis in the preceding section (such as DFKI_dk, IMS_de, KTH_se and COGSCIED_uk). It is also noteworthy that some institutions, even if not being in the restricted group of the very best-connected, are quite "visible" from the outside of the network. This is the case of both research groups from the University of Cambridge, the Computer Laboratory and the Speech, Vision and Robotics Group and the Universitaet des Saarlandes Department of Computer Linguistics.

Moreover, there are a certain number of institutions that, though they do not belong to the group of electronically best-connected ones within the ELSnet network, do have a very high degree of "visibility" in relation to the external network. This is important for they might be considered as "authoritative"

Table 6. Who is directly linking to one's institution?

Direct InLinks to sample of best connected and most "prestigious"					
	Percentage of Direct InLinks (by type)				
Research Institutions	*"Inbreeding"*	*Best Connected*	*Other ELSnet*	*Hubs /Authorit*	*Other*
LIMSI/CNRS - Human-Machine Communication Department	58,6	4,2	7,4	2,4	27,3
German Research Center for Artificial Intelligence (DFKI) - Language Technology Lab	55,0	11,3	4,6	6,6	22,5
Universitaet des Saarlandes Department of Computer Linguistics	57,8	10,3	5,3	3,2	23,4
Universität Stuttgart-IMS - Institut für Maschinelle Sprachverarbeitung	26,5	11,9	8,8	7,5	45,2
KTH (Royal Institute of Technology) - Department of Speech, Music and Hearing (TMH)	48,5	4,2	3,3	3,9	40,1
University of Cambridge - Computer Laboratory	50,3	2,5	2,0	2,4	42,9
University of Edinburgh - Informatics	26,4	9,4	5,1	6,3	52,8
University College London - Department of Phonetics and Linguistics	56,0	3,1	4,2	3,5	33,2

resources from outside the boundaries of ELSnet. This is the case of the Université Paul Sabatier (Toulouse III)—Institut de Recherche en Informatique de Toulouse (IRIT), Xerox Research Centre Europe—Grenoble Laboratory, Langenscheidt KG—Electronic Media from Germany, Universität des Saarlandes CS-AI—AI Lab at the Department of Computer Science and the Institute for Language & Speech Processing (ILSP) in Greece. This result is also important for the discussion of modeling the discovery of digital knowledge bases, as this indicates that not only should the core group of densely connected institutions be extensively mapped, but also some of the less well connected within the network, which are nonetheless "authoritative" in terms of external links.

These results led to a more detailed analysis of "Who is linking directly to one's institutional electronic site?" An analysis of a sample of the electronically best-connected institutions as well as some of the ones with more direct inlinks provided some interesting results, summarized in Table 6.

A very strong conclusion from these results is the high percentage of self-citation or direct links coming from inside the research institution's home pages. This is an indicator of electronic "inbreeding." At least 50% to 60% of the total number of links connecting to one institution is coming from inside the institution. This is valid for all but two cases, that are the most very well connected institutions (the Universitat Stuttgart—IMS and the University of Edinburgh—Informatics). It comes as no surprise that the best-connected institutions should also be the less "inward looking" institutions. But it is highly satisfactory to confirm this conceptual assumption. It is also noteworthy that a significant proportion of the self-citation links refer to the institutional internet resources of staff information and project information. This is highly relevant for the discovery of electronic knowledge bases (to be discussed in the next section).

As we have focused our analysis on the best electronically connected institutions and the most "central" ones, it comes as no surprise that the percentage of direct inlinks coming from other ELSnet best-connected institutions is significant and tends to be higher for the very best-connected ones. Another interesting finding that reinforces the results from the preceding sections is that the very best-connected institutions tend to directly cite each other. They indeed form a dense subgroup of highly connected institutions, revealing a kind of "prestige accumulation process." This process is again reinforced by the relatively high percentage of external entities (hubs, authorities and internet resources in the field) that link directly to those institutions. It is also important to stress the direct connectivity from other member institutions of the ELSnet, which is in accordance with the idea that this core set of institutions function as "internal authorities" to the network. A final comment on the results is that there is a significant number of links coming from other miscellaneous institutions (a percentage around 30%-45%), some of which are related to the field of research and development but some of which are not. This external connectivity is important for generating heterogeneity in the knowledge discovery process and opening the "digital knowledge base" to external sources (see following sections).

Measuring the Electronic Activity of Research Institutions: Out-Degree-Centrality

Another complementary measure of the centrality of institutions within a network is given by the outdegree centrality, or number of links originating in a certain institution. Given the hyperlinked nature of the internet, this indicator is not measurable, unless we restrict the *depth* of the outgoing links to a certain n degrees of connectivity. By doing this we are able to compute and visualize the ego-centered network of any institution, to a depth n of analysis.

With some similarity to the previous analysis of direct incoming links to a certain institution, the analysis of outgoing links also reveals important characteristics of the connectivity of one's institution.

From the analysis of the electronic ego-networks of a restricted sample of the most well-connected institutions, the results confirm interesting expectations:

- The proportion of outgoing links to information resources within the same institution is quite significant. So, here again we have a significant percentage of "inbreeding." And again this is highly correlated with information on staff and projects, as well as internal "knowledge-bases," for example, chapters, reports, and so on.

Figure 2. Mapping individual research institutions' electronic ego-networks

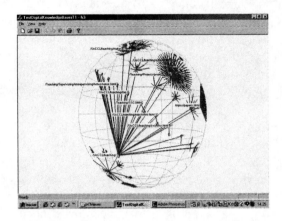

- We can clearly identify strong connectivity with those institutions with which one's institution collaborates more intensively (particularly in research projects and in the inter-exchange of research staff, such as in PhD or postdoctoral work).
- We can identify external linkages to important authoritative resources.

Figure 2 shows the results for the Computer Laboratory (University of Cambridge). The ego-network electronic map reveals a concentration of some important information resources within the institution. Those "clusters" of information include chapters published by the researchers, online teaching resources as well as projects undertaken by the research groups. At a more distanced "layer" of connectivity (*n* greater than five) we could also map links to external authoritative resources, as well as links to other members of the research network.

A Conceptual Model for the Discovery of Digital Knowledge Bases

This section is exploratory in nature, as we have limited empirical evidence given by the reduced sample of five electronic ego-networks analyzed. Nevertheless, the results from the preceding two sections are quite revealing of the importance of analysing the topology (link-structure) of research networks in order to gain a better understanding of the structure of the collaboration between those institutions, as well as the information resources those institutions make available in electronic space.

From the outset we should stress the fact that this analysis is generally applicable only to those kinds of institutions that openly disclose their information and "knowledge-resources" in a publicly accessible electronic form. In a previous section (see patterns of electronic connectivity above), the empirical data reveal that private companies are not likely to openly and publicly disclose their information online. As such, we have to take into consideration that these results are only more generally applicable to research institutions from the university and government laboratory sectors.

Among this subset of research institutions that are members of a known network of collaboration in the field of research, we are likely to find a reasonable heterogeneity in terms of patterns of connectivity, as well as centrality and prestige within the network. These results have been discussed in the preceding two sections. The empirical evidence validates the existence of subgroups within an *inner structure* of these electronic networks that are more strongly connected with one another, as well as maintain a reasonable percentage of links with other members of the nonelectronic network, and links to external "authoritative" resources. The detailed examination of some of these very-well-connected and central institutions, has revealed disclosure in electronic webspace of important knowledge resources, such as information on publications in digital form, research project documentation and research staff information.

By extensively mapping these ego-centered networks in electronic webspace, we are able to detect "overlapping regions" where information resources from these institutions are linked together, for example, the Computer Laboratory with the Cognitive Sciences Department of the University of Sussex, the Computational Linguistics Department of the University of Sheffield, the University of Nijmegen with the University of Stuttgart—IMS and the University of Sheffield.

A more general procedure for detecting very intensive electronic zones of knowledge-resources, related to a specific field of research, is potentially described in the following way:

- Start with a well-defined set of institutions that by their collaboration activity reveal a structural pattern of connectivity—this is usually the case with research networks;
- Collect information about the electronic centrality of all those institutions, particularly the number and type of direct in-links;
- Map the electronic ego-centered networks of all those institutions, to a certain "layer" of connectivity, at least 10 links in depth;
- More selectively, map only those electronic linkage structures for the institutions revealed to be best-connected as well as more central and "prestigious";

- Detect the "overlapping regions" of connectivity between the whole set of ego-networks, as well as the most relevant external linkages linking to/from members of the network (good candidates for external "authorities" and "hubs"); and
- Map the ego-centered network of those external "hubs" and "authorities."

This will allow one to identify digital knowledge bases, or in other words, zones of the electronic networks that are very intensive in information resources, and are simultaneously linked to important members of the scientific research network.

Two important characteristics of this methodology are worth commenting upon. First, the initial restriction to a set of collaborating institutions is crucial, in terms of the precision of the analysis. In fact, by limiting the analysis to research institutions that are likely to collaborate more intensively, we are likely to restrict the content nature of the results to knowledge resources relevant to the field of research under investigation. Secondly, the integration into the overall analysis of important "external" authoritative institutions and resources brings diversity to the initial network connections, as well as potentially allowing the extension of these digital knowledge bases to a broader spectrum not foreseen at the beginning. This heterogeneity gain is likely to yield important results in terms of the quality of information one can access within these digital knowledge bases.

The conceptual model represented in Figure 4 allows the identification of the essential elements:

a. the electronic ego-networks for the best-connected institutions (linkages to a depth n of connectivity);

b. the cross-linkages between research institutions' ego-networks at any "layer" level and the electronic connectivity to external authoritative resources;

c. the in-links coming from hubs, authorities and external internet resources to best-connected institutions;

d. the identification of network zones of intensive information resources— digital knowledge bases, resulting from the connectivity of the institutions.

An integrated representation of these concepts and processes is shown in Figure 3.

Notwithstanding the potential identification of these digital knowledge bases, there are two important limiting factors to be taken into account:

Figure 3. Identification of digital knowledge bases

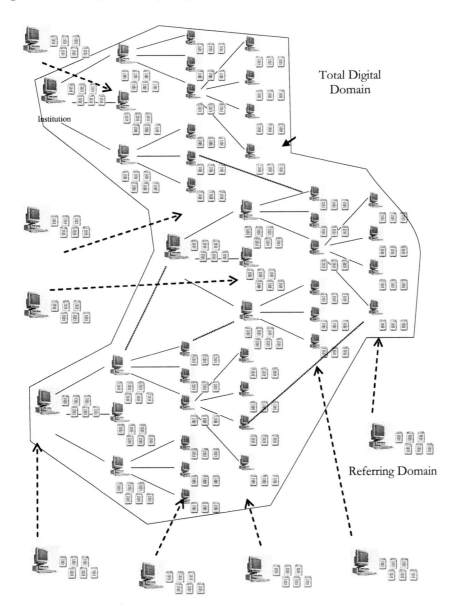

Total Digital
Domain

Institution

Referring Domain

Legend:

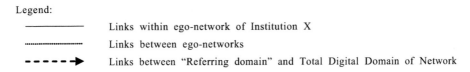

—————————— Links within ego-network of Institution X

.... Links between ego-networks

- - - - ➤ Links between "Referring domain" and Total Digital Domain of Network

- There is no *refereeing* mechanism or scientific validation of the "knowledge" resources available electronically, other than the self-organization process inherent to the institutional practices of disclosing information on these electronic networks. This might represent a problem in terms of quality of the digital knowledge base resources;

- This analysis is cross-sectional in nature and does not account for the dynamic nature of these electronic systems. The structural representation of the electronic connectivity and the mapping of knowledge bases constitute a "snapshot" taken at a certain period in time. But empirical evidence testifies to the changing nature of information published on the web, which might cause dramatic changes in these structural representations. This problem can only be overcome by a longitudinal analysis of these networks.

Conclusion

Recent research has highlighted the existence of mechanisms of "self-organization" of large, distributed electronic networks such as the internet, and the efficient identification of "web communities" in digital spaces. In this research we have focused our analysis on the *inner structure* of such "web communities," namely the connectivity patterns and centrality of certain institutions within these restricted networks of electronic interactivity. Moreover, as the analysis is particularly concerned with scientific communities and research networks, the sociological context embedding the collaboration structures is likely to contribute to a large degree to explain the overall system. The argument is that the electronic connectivity patterns "extend" previous and more or less well-established collaboration patterns in the offline world, and not the other way around.

We have found that indeed, the pattern of electronic connectivity of institutions belonging to research networks shows a remarkable similarity with offline, nonelectronic collaboration structures. Moreover, the "best-connected" group of institutions supports a kind of self-reinforcing mechanism assuring the density of their inner connections. Secondly, we have detected heterogeneity in the *inner structure* of the electronic networks, even among the "best-connected" entities. Therefore, there is a reduced subgroup of "very best"-connected institutions not only within the "internal" network under analysis, but also with important external connections to "authoritative" institutions and internet resources. For the institutions that have an open and public policy of information disclosure within these electronic networks (and this excludes most of the private business companies), the extensive mapping of their digital ego-networks allows one to identify important zones of information and knowledge resources—

"digital knowledge bases." These electronic resources (e.g., electronic publications, projects documentation, researcher's information) are linked to the best-connected institutions in the research network, as well as to important hubs and authorities in electronic webspace, as described above.

Science Policy Implications

The findings discussed above have important implications for science policy, with a particular emphasis in three dimensions. First, the availability of knowledge resources in electronic networks and infrastructures for knowledge production and distribution. Secondly, the organization of research networks and their production and distribution of knowledge. Finally, the opportunity of new and innovative forms for science policy assessment and evaluation.

Electronic networks provide new opportunities and infrastructures for knowledge creation and the dissemination of knowledge. Nevertheless, these new infrastructures represent a potentiality, and its realization is significantly dependent upon the effective existence of knowledge resources widely available on the World Wide Web and their subsequent use in collaborative environments by different communities of practice. The identification of "digital knowledge bases" as described in this investigation, allows the delimitation of intensive knowledge resources connected on the web with a topical nature, which represents a necessary but not sufficient condition for this realization. Moreover, the clear anticipation of certain forms of resources availability in these wide and open electronic networks provides essential guidelines informing the planning, design and implementation of new infrastructures for knowledge distribution and use (despite the crucial need to take into account the information-rich context and variety in cultures embedding ICT infrastructure policies—see Merz in this volume).

The networking of research entities and research resources creates new possibilities for knowledge creation. Digital knowledge bases can provide a better understanding of particularly effective forms of organizations of research networks. Here, and again, this should not mean automatic forms of reproduction of differentiated cultures and practices into the digital realm (see Fry in this volume for discussion of the cognitive and social shaping of collaboration). Nevertheless, the exploration of different organizational structures, prior to their actual implementation, has the potential to offer significant cost reductions, as well as science policy foresight. Additionally, the empirical verification of "digital knowledge bases" provides a more dynamic process for new and innovative forms of research organization leading to network structures less dependent on physical infrastructures and more flexible and adaptable to science policy transformations.

We are still at an early stage in understanding how particular forms of origination lead to certain levels of research performance. The causality between structure and origination on the one hand and functionality or performance on the other is not yet clearly understood. This argument is valid for the various levels of research assessment (individual, departmental, institutional, or international). Nevertheless, it looks worthwhile to explore the validation of offline and more traditional methods for research evaluation (based upon publications, projects, expert assessment, and other scientometric indicators) with online indicators (online centrality, prestige, digital knowledge bases, and other webmetric indicators). The coming decades will provide more clear insight into the actual mapping of offline structures and their online representations and vice-versa. This is likely to provide better and more informed methods for research evaluation.

Avenues for Further Research

A serious limitation to this investigation is the cross-sectional character of the research. Indeed, a longitudinal analysis of the dynamics of these connectivity patterns, as well as of the corresponding dynamics of the digital knowledge bases, is an important line for further empirical investigation. A comparison of the same conceptual scheme in different research networks, for example in different fields of research, or in research networks resultant from different funding policies, is another important area for further analysis. Discussion of webmetric indicators, at the micro-level of single institutions, the meso-level of groups of institutions or the macro-level of whole networks, is also a research topic in itself, as is research directed to analyze the policy implications of the commercial practice of non-public disclosure of information in these electronic environments. The same is true for copyright issues related with the information actually being disclosed by university and governmental institutions. The technological and sociological dimensions of the problem of public availability of knowledge resources in electronic networks will potentially advance side by side in order to facilitate the positive prospects of such communication systems, and fully exploit the creation of "webs of knowledge" on the internet.

References

Albert, R., Jeong, H., & Barabási, A-L. (1999). Diameter of the World Wide Web. *Nature, 401,* 130-131.

Aquino, S., & Mitchell, T. (2001). Search engines ready to learn. *Technology Review Online,* April 24.

Bordóns, M., Zulueta, A., Cabrero, A., & Barrigón, S. (1995). Identifying research teams with bibliometric tools. In *Proceedings of the Fifth Biennial Conference of the International Society for Scientometrics and Informetrics* (pp. 83-92). Medford, NJ: Learned Information.

Brin, S., & Page, L. (1998). The anatomy of a large-scale hypertextual web search engine. In *Proceedings of the Seventh International World Wide Web Conference* (pp. 107-117). Amsterdam: Elsevier.

Caldas, A. (2004). *The structure of electronic scientific communication: Electronic networks, research collaboration and the discovery of digital knowledge bases.* Unpublished doctoral dissertation, University of Sussex (SPRU).

Cooley, R., Mobasher, B., & Srivastava, J. (1997). Web mining information and pattern discovery on the World Wide Web. In *Proceedings of the 9th International Conference on Tools with Artificial Intelligence (ICTAI 97)* (pp. 558-567).

Flake, G., Lawrence, S., & Giles, G. (2000, August 20-23). Efficient identification of web communities. In *Proceedings of the Sixth International Conference on Knowledge Discovery and Data Mining* (ACM SIGKDD-2000) (pp. 150-160). Boston: ACM Press.

Flake, G., Lawrence, S., Giles, C., & Coetzee, F. (2002). Self-organisation and identification of web communities. *Computer Magazine,* March, 66-71.

Garfield, E. (1979). *Citation indexing: Its theory and application in science.* New York: Wiley.

Garton, L., Haythornthwaite, C., & Wellman, B. (1999). *Studying online social networks.* In S. Jones (Ed.), *Doing internet research* (pp. 75-105). Thousand Oaks, CA: Sage.

Huang, L. (2000). *A survey on web information retrieval technologies.* Retrieved June 6, 2002, from State University of New York at Stony Brook web site: http://citeseer.nj.nec.com/336617.html

Huberman, B., Pirolli, P., Pitkow, J., & Lukose, R. (1998). Strong regularities in world wide web surfing. *Science, 280,* 95-97.

Katz, S., & Martin, B. (1997). What is research collaboration? *Research Policy, 26*(1), 1-18.

Kleinberg, J. (1998). Authoritative sources in a hyperlinked environment. In *Proceedings of the acm-siam symposium on discrete algorithms* (pp. 668-677).

Kleinberg, J., & Lawrence, S. (2001). The structure of the web. *Science, 294,* 1849-1850.

Koku, E., N. Nazer, & Wellman, B. (2000). Netting scholars: Online and offline. *American Behavioral Scientist* [Special Issue], "Mapping Globalisation," *43*(10), 1752-1774.

Kosala, R., & Blockeel, H. (2000). Web mining research: A survey. *Proceedings of ACM—sigkdd explorations* (pp.1-15).

Kumar, R., Raghavan, P., Rajagopalan, S., & Sivakumar, D. (2000). The web as a graph. *Proceedings of the ACM-sigmod-sigact-sigart symposium on principles of database systems* (pp. 1-10).

Kumar, R., Raghavan, P., Rajagopalan, S., & A. Tomkins (1999). Extracting large-scale knowledge bases from the web. *Proceedings of the 25th very large data bases conference*, Edinburgh, Scotland (pp. 639-650).

Larson, R. (1996). Bibliometrics of the world wide web: An exploratory analysis of the intellectual structure of cyberspace. *Proceedings of Annual Meeting of the American Society for Information Sciences* (pp. 71-78).

Lawrence, S., & Giles, C. (1999). Accessibility of information on the web. *Nature, 400*(6740), 107-109.

Moreno, J. (1932). *Group methods and group psychotherapy* [Monograph]. Beacon: Beacon House

SearchEngine ShowDown (2002). *Statistics for the size of the web.* Retrieved March 31, 2005, from   http://www.searchengineshowdown.com/stats/size.shtml

Van Raan, A. (1998). Assessment of social sciences: The use of advanced bibliometric methods as a necessary complement of peer review. *Research Evaluation, 7*(1), 2-6.

Wasserman, S., & Faust, K. (1994). *Social network analysis: Methods and applications.* Cambridge: Cambridge University Press.

Wellman, B. (1996). *For a social network analysis of computer networks: A sociological perspective on collaborative work and virtual community.* SIGCPR/SIGMIS '96, Denver Colorado, USA, ACM.

Endnote

[1]   At first, the connectivity result for the University of Edinburgh was somewhat surprising. However closer inspection revealed that this node contains a cluster of three very respectable research groups in this field: the Centre for Speech Technology Research, the Linguistics Department and the Cognitive Sciences research group.

Chapter XI

From the "Analogue Divide" to the "Hybrid Divide": The Internet Does Not Ensure Equality of Access to Information in Science

Franz Barjak
University of Applied Sciences Solothurn Northwestern Switzerland,
Switzerland

ABSTRACT

This chapter investigates whether the internet has improved information access for scientists who did not participate fully in the transfer of information in pre-internet times. Several empirical analyses over the last decade have nurtured the hope that the internet had this effect. We argue that these findings were mostly due to the low level of dissemination of the internet in the early 90s. Based on a large European data set, we show that internet use is consistently higher for male, highly recognized and senior researchers. This suggests that the internet has become the dominant means of communication in science—to such an extent that any scientist, regardless of whether they are established or not, has to use the available internet tools in order to communicate effectively. The previous "analogue divide" of information access has become a "hybrid divide" including the analogue and the digital communication media.

Introduction

The OECD (2001) defines the catchword "digital divide" as "the gap between individuals, households, businesses and geographic areas at different socioeconomic levels with regard both to their opportunities to access information and communication technologies (ICTs) and to their use of the internet for a wide variety of activities" (p. 5). This gap has been identified for several socioeconomic groups, like people with disabilities, people on low incomes and from households with certain features (e.g., single parent households), the unemployed, people with relatively low levels of skills and educational attainment, people with literacy difficulties, people belonging to racial and ethnic minority groups, people living in remote rural locations and, last but not least, senior citizens (see, e.g., National Telecommunications and Information Administration [NTIA] & Economic and Statistics Administration [ESA], 2000; OECD, 2001; Work Research Centre, 2003).

In analogy with this concept we use the term "analogue divide" to describe the disparate use of information sources and information access in pre-internet times. In science, this analogue divide was pronounced before the internet spread: Scientists that were outside or at the fringe of an invisible college[1] did not receive as much information as its participants, and they obtained information with delays (Cole & Cole, 1973; Crane, 1972; Cronin, 1982; Mulkay, 1977). The system of information exchange discriminated against scientists that were younger, female, at lower positions in the hierarchy of their research organization, of lower professional recognition, working at less renowned universities or in developing countries.

The spread of the internet has raised hopes that information access for those disadvantaged groups would improve and that the internet would contribute to creating a more equal dissemination of information and communication in science (Hilgartner, 1995; Finholt & Olson, 1997; Walsh & Roselle, 1999). In this chapter we investigate whether these hopes have turned into reality. Most empirical analyses over the last decade have revealed more intensive internet use by less established groups of scientists. The data analyzed in this chapter do not support this result: On the contrary, internet use is consistently higher for the scientific establishment. We argue that the internet has become the dominant means of communication in science—to such an extent that any scientist, regardless of their position and status, has to use the available internet tools and applications in order to communicate effectively. There are, as we will also show, some differences of internet use between scientific disciplines and countries, but no particular benefit can be found for less-established scientists.

Current State of Knowledge

There are convincing arguments for and against the internet's positive contribution to closing the information gap between established scientists and their less acknowledged colleagues (see on this issue Nentwich, 2003): On the one hand the internet has improved access to information from scientific journals, databases, archives, and other information sources; scientists don't need to obtain one of the rare journal issues, they don't have to be on the mailing lists of eminent colleagues to receive their latest results and they need to invest less of their own or their assistants' time for literature searches. Furthermore, e-mail provides a direct gateway to other scientists without having to reveal one's identity and status (Dubrovsky, Kiesler, & Sethna, 1991; Walsh & Roselle, 1999). On the other hand, access to electronic information sources is often not free of charge. It requires costly subscriptions and licences and may be restricted to a limited number of work stations. In regard to personal communication, overloaded scientists might have their e-mail screened and filtered. The filtering of online communication partners is facilitated by status cues such as addresses, biographies and photos which are increasingly inserted into WWW sites (Nentwich, 2003; Walsh & Roselle, 1999).

If the internet had contributed to reducing the analogue divide in terms of information access and use of information sources in science, we would expect that groups which were formerly disadvantaged—like younger, female, less renowned and geographically peripheral scientists—would use the internet more, or at least to the same extent, as their more established peers.

Empirical analyses over the last decade have largely confirmed this assumption: Lazinger, Bar-Ilan, and Peritz (1997) obtained lower internet use for more senior scientists; correspondingly, Cohen (1996) and Mitra, Hazen, LaFrance, and Rogan (1999) established that younger scientists used computer-mediated communication (CMC) applications more often. Kling and Callahan (2002) found that e-journal readers tended to be younger than nonreaders. A "lower" professorial status (assistant versus associate and full professors) also corresponded to a higher use of computer networks for communication (Cohen, 1996). Only the findings on gender differences between scientific CMC users were inconclusive: Whereas Cohen (1996) reported higher CMC use by women, Mitra, et al. (1999) and Walsh, Kucker, Maloney, and Gabbay (2000) did not find any gender-related differences in regard to e-mail use. However, these results are mostly based on data that were collected when the internet was still in its infancy. More recent data are necessary to obtain a picture of the current reality.

Internet Access and Use for Different Status Groups in Science

The Statistical Indicators for Benchmarking the Information Society (SIBIS)[2] survey collected data on more than 1,400 scientists from five academic disciplines (astronomy, chemistry, computer science, economics, and psychology) and seven European countries (Denmark, Germany, Ireland, Italy, the Netherlands, Switzerland and the UK). The data were gathered through a mailed paper-and-pencil questionnaire in the period between April and July 2003 (response rate: 25%). It permits a comparison to be made of scientists' internet access and use according to age, gender, position in regard to R&D and academic recognition.

E-mail is used by 99.7% and the World Wide Web by 98.9 % of the respondents to the survey, and we can therefore state that virtually every respondent uses the internet. It was possible to analyze three aspects of information access and use of information sources for R&D with the data set:

a.   Internet access to important information sources;

b.   Use of online and offline information sources; and

c.   Use of communication media in an average working week.

The results are shown in the Tables 1-3 and will be discussed in turn.

a.   Internet access to important information sources (Table 1): Overall, internet-based access to information sources which the respondents consider as important from their point of view is very good. On average, only 3.5% of all respondents stated that they had access to only few or none of their important information sources. However, access varies according to academic discipline, country and between the different status groups. Higher percentages of female, younger and less-renowned scientists could

Table 1. Percentage of scientists with insufficient internet access to the important information sources and reasons given for this insufficient access by status group, academic discipline and country (SIBIS survey on the internet in R&D)

| | | Insufficient internet access to the important information sources | Reasons given for this insufficient access[b] | | |
			Information source is not available on the internet	Technical infrastructure is insufficient	Organization has not enabled access
Gender	Male	3.3	34.3	11.8	51.4
	Female	4.4	16.7	16.7	61.1
Age	35 and younger	4.0	30.0	5.0	71.4
	36 to 50	3.7	25.0	15.8	57.9
	51 and older	2.8	30.8	23.1	23.1
Level of recognition	Low	3.8	27.0	11.1	64.9
	Medium	3.7	36.4	18.2	18.2
	High	2.3	a	a	a
Position	Research Manager	3.6	30.0	10.0	50.0
	Senior Researcher	3.2	34.8	18.2	36.4
	Junior Researcher	3.3	20.0	6.7	81.3
Academic discipline	Astronomy	1.1	a	a	a
	Chemistry	2.9	a	a	a
	Computer Science	3.5	0.0	30.0	50.0
	Psychology	5.2	46.7	7.1	35.7
	Economics	3.3	a	a	a
Country	Switzerland	1.3	a	a	a
	Germany	2.2	a	a	a
	Denmark	5.6	a	a	a
	Italy	4.0	25.0	33.3	41.7
	Ireland	7.2	9.1	18.2	81.8
	The Netherlands	3.5	a	a	a
	United Kingdom	2.9	a	a	a
All respondents		3.5	28.3	13.5	54.7

[a] Less than 10 observations and therefore excluded from the table. [b] Only for respondents that state insufficient access; percentages don't add to 100 because multiple responses were possible and the answering category "other reasons" is not shown.

not access the important information sources via the internet. The major reason in each case was that their organization has not enabled the access. This reason is important primarily for the lower status groups, whereas insufficient technical infrastructure and the general absence of these sources from the web are less important.

Among the disciplines included in the survey, astronomers are the scientists with the best internet access to information sources. Psychologists have the worst access, but still only 5% state that it is insufficient. Country differences are more notable. In Ireland, in particular, the share of respondents that cannot access the important information sources is twice as big as the average, and the most important reason for this is that their organization has not enabled access.

b.   Use of online and offline information sources (Table 2): Most internet-based information sources shown in Table 2 are used to a large extent. In particular, electronic journals and databases, library web sites and web sites of other institutions are used each by more than 60% of the respondents. Among the offline sources, the scientists' own collections of information and coworkers are particularly popular. Male scientists are slightly more regular users of information sources than female scientists.[3] Scientists in higher positions (research managers and senior researchers) and with higher professional recognition[4] are also more often regular users of online sources than colleagues in lower positions (junior researcher) or with less recognition. This is particularly notable for peer web sites and web pages of other organizations, but it also applies to electronic journals.[5] Only the age differences are in line with the expectation that less-established scientists use online sources more frequently than established scientists: Younger scientists are more often regular users of openly accessible information sources such as internet sites of libraries, electronic journals and databases than older scientists, whereas they are less often regular users of their own collection of information items, offline electronic sources and conferences, workshops and seminars.

A specific pattern of sourcing information appears for each scientific discipline included in the analysis. Psychologists use all information sources less than the average. Economists rely in particular on internet-based sources. Astronomers source information through social interaction (co-workers, conferences, workshops) and from internet-based databases. Chemists rely on library-related sources (online and offline) and computer scientists on peers' and other institutions' web pages. The differences between countries are less pronounced than between scientific disciplines. Irish and British scientists tend to use all information sources less than their peers from the other countries, whereas Swiss and Italian scientists tend to use them more.

c.   Use of communication media in an average working week: Based on an assessment of the use of communication media during an average working week, three clusters were identified in a hierarchical cluster analysis:[6]

- "Silent researchers" use all the listed communication channels only to a small extent.

- "E-mail communicators" use mainly e-mail for their R&D communication.

- "Communicators" use all the different communication media intensively.

Table 2. Percentage of regular users (once a week and more) of an information source by status group, academic discipline and country (SIBIS survey on the internet in R&D)

		Internet sites of libraries and archives	Electronic journals, working papers and article databases	Peers' web pages	Websites of other institutions	Your own collection of information items	Off-line electronic sources	Libraries	Colleagues, assistants, superiors	Conferences, workshops, seminars
Gender	Male	68.3	76.5	45.0	65.5	82.9	42.1	43.2	64.4	18.1
	Female	67.6	76.6	31.8	52.9	79.5	33.1	39.2	57.6	13.9
Age	35 and younger	71.6	83.2	39.9	54.0	79.7	34.6	40.1	65.5	14.7
	36 to 50	67.6	75.3	45.7	67.7	82.1	38.9	36.3	61.3	15.5
	51 and older	64.4	70.2	39.7	66.0	85.4	47.9	52.2	61.2	21.8
Level of recognition	Low	67.1	76.0	39.1	59.9	80.9	37.4	42.2	60.8	14.4
	Medium	68.8	81.5	45.0	68.4	81.6	39.1	41.3	66.5	21.0
	High	71.7	73.6	51.4	68.1	88.8	52.6	44.7	67.6	24.2
Position	Research Manager	65.5	75.2	50.4	70.6	84.2	45.1	41.4	63.8	23.0
	Senior Researcher	69.4	78.8	44.1	66.3	84.4	40.6	43.9	61.5	17.8
	Junior Researcher	69.5	77.3	35.8	54.0	79.5	37.7	41.5	65.3	13.9
Academic discipline	Astronomy	73.2	89.5	36.6	79.5	80.6	45.7	42.0	76.9	23.0
	Chemistry	77.0	88.9	34.8	46.5	82.7	42.3	52.7	59.1	15.4
	Computer Science	54.7	60.3	64.8	73.2	82.4	33.2	27.3	65.8	17.8
	Psychology	70.8	67.0	30.7	49.4	82.8	40.7	45.1	57.7	9.4
	Economics	70.3	83.5	47.8	70.7	83.3	39.7	42.9	59.7	22.5
Country	Switzerland	66.7	78.7	46.9	71.6	81.1	43.6	40.6	70.7	20.4
	Germany	67.6	70.8	46.3	69.2	82.8	41.3	50.9	63.0	19.8
	Denmark	66.5	75.0	41.0	61.2	86.8	46.7	42.6	66.3	16.0
	Italy	74.0	82.1	43.7	66.2	78.1	47.1	44.3	58.7	18.8
	Ireland	64.2	73.5	34.7	50.0	85.1	30.8	40.9	51.7	8.1
	The Netherlands	71.3	81.0	33.6	50.7	81.2	35.2	35.5	70.9	15.1
	United Kingdom	63.6	73.9	39.5	55.4	82.4	26.2	35.4	57.0	16.7
All respondents		68.1	76.6	42.0	62.6	82.1	40.0	42.4	62.7	17.1

How the different status groups are represented in these communication clusters is shown in Table 3. For our purposes, the second cluster of "E-mail communicators" is the most interesting, as we would expect that the disadvantaged groups of scientists are overrepresented in this cluster. However, this is hardly the case. Only the youngest age group, scientists of 35 years and younger, is slightly overrepresented compared to scientists of 51 years and older. The other low status groups, female, little recognised and lower-positioned scientists, are always clearly overrepresented in the "Silent researchers" and underrepresented in the "Communicators" clusters.

Somewhat in line with their geek image, computer scientists are overrepresented in the cluster of e-mail communicators. So are astronomers who overall communicate more, probably as astronomy is a collaboration-intensive science (Barjak & Harabi, 2004; Crawford, Hurd, & Weller 1996). Chemists are overrepresented among the silent researchers. This also applies to Irish scientists. Scientists from the UK are more often to be found among the e-mail communicators, and scientists from Italy and Germany among the communicators cluster.

Table 3. Percentage of respondents in clusters of communication channel usage by status group, academic discipline and country (SIBIS survey on the internet in R&D)

		Silent researchers	E-mail communicators	Communicators	Total
Gender	Male	39.1	28.2	32.7	100.0
	Female	49.3	27.7	23.0	100.0
	35 and younger	52.9	29.3	17.8	100.0
Age	36 to 50	34.2	31.0	34.8	100.0
	51 and older	37.5	22.8	39.7	100.0
Level of	Low	51.6	26.6	21.7	100.0
recognition	Medium	23.9	32.1	44.0	100.0
	High	15.1	31.2	53.7	100.0
	Research Manager	21.4	25.5	53.1	100.0
Position	Senior Researcher	32.5	32.5	34.9	100.0
	Junior Researcher	62.3	25.3	12.4	100.0
	Astronomy	32.4	33.0	34.6	100.0
Academic	Chemistry	54.6	15.5	29.9	100.0
discipline	Computer Science	36.2	37.8	26.0	100.0
	Psychology	36.2	30.6	33.2	100.0
	Economics	44.4	27.3	28.3	100.0
	Switzerland	43.9	34.1	22.0	100.0
	Germany	40.9	19.2	39.9	100.0
	Denmark	41.1	35.0	23.9	100.0
Country	Italy	32.4	23.4	44.1	100.0
	Ireland	56.8	23.3	19.9	100.0
	The Netherlands	45.7	28.6	25.7	100.0
	United Kingdom	37.9	39.1	23.0	100.0
All respondents		41.5	28.2	30.4	100.0

Discussion and Conclusion

The SIBIS survey data do not support the hypothesis that differences in access to information between lower and higher status groups are overcome by means of the internet. On the contrary, higher status groups, with the exception of older scientists, have consistently better access to internet-based information sources. They are more often regular users of online information sources and they are more intensive users of e-mail as well as most other communication channels. Hence, the analogue divide in information access between established and not established groups in science has been carried over to the internet age. The internet does not offer a particular advantage for the previously discriminated against groups. It seems that the mechanisms that caused the analogue divide, such as disparate financial resources for journal subscriptions, other published information and conference attendances, differing resources (above all personal assistance) for searching for and filtering information and unequal participation at academic in-groups are still valid. To some extent there are even online equivalents of these access barriers. It would appear that the old analogue divide has become a new "*hybrid divide*" including analogue and digital information and communication.

How can this be reconciled with the results of older analyses? The latter mostly date back to the early phase of the diffusion of the internet in science. When the internet was only one of several communication media, established scientists could still rely on other media and ways of communicating. At that time, discriminated groups may have had an advantage by using the internet to access material that they could not obtain offline. Nowadays, at the advanced stage of internet diffusion, it has become the dominant means of information retrieval and communication. Though field differences persist, no scientist can afford to ignore the internet entirely. Established scientists must use it in order to work effectively. Partially, they also have the means to regulate and control internet access and the use of online information sources by less-established scientists. The situation is therefore the same as before: It is the hierarchy and power distribution in science, not the technology, which determines access to information. This finding strengthens propositions that the internet does not change the social order in science (Gläser, 2003); instead, it is integrated into scientific knowledge production in a way that leaves the existing social order intact.

Some qualifications are, however, necessary. First of all, the dataset was limited to scientists from seven developed European countries. The internet might have had more notable effects on information access for scientists from developing countries. Also, the reputation, research orientation and location of the organization could not be included among the discriminatory variables. Second, the data merely cover use of internet-based information sources and communication channels. They do not measure the significance and value of the information obtained from online sources. However, this information might have a higher value for the less-established scientists, as they could not obtain it by other means and don't have any substitutes. Third, younger scientists are more often regular users of internet-based information sources than older scientists. This difference might be due to the main hypothesis of this paper. However, as none of the other status variables support the hypothesis, other explanations come to mind. It might not be their lower status, but other reasons that motivate younger scientists to use internet-based information sources. For instance, we can assume that younger scientists show more affinity to new technologies and have higher incentives to learn how to use them, as they can expect higher cumulated returns during their future career. All in all, however, we cannot negate the fact that the internet might have solved some of younger scientists' problems of obtaining information. Therefore, we may conclude that up to now the internet has not bridged the information divide in science entirely; but it might have reduced it at least in part.

The analysis has also shown some differences of internet use between scientific disciplines and countries. The disciplinary differences are not really investigated in the present chapter. Previous research suggests that disciplines are still rather heterogeneous and that the field level is more appropriate for analyzing ICT use (Fry, 2004 and this volume; Kling & McKim, 2000). They point to several factors such as the work products, the work organization and institutional factors. However, we find these factors reflected in our results: Astronomical research often takes place in ICT-supported collaborations (Barjak & Harabi, 2004; Crawford, Hurd, & Weller 1996), and the astronomers in the survey source information a lot through social interaction and from electronic databases. Chemists rely especially on library-related information sources; also, e-mail doesn't seem to play the key role in their communication activities. This is in line with their work pattern in research that includes time spent in the laboratory and away from the desk. Kling and McKim (2000) reported that in computing, scholarly societies adopted a much more open stance towards electronic pre- and re-publications of papers than, for instance, in psychology (at least in the United States). We find that European computer scientists use the computer more than the other scientists as a communication medium and they source a big part of their information needs from the web. European psychologists, on the other hand, make relatively little use of electronic journals and other online information sites. However, the level of disciplines chosen for this analysis probably masks some of the more fine-grained differences between fields within a discipline.

In regard to the country pattern, the lack of internet use by Irish scientists is striking. However, it is not only a lack of use of the internet, but a lower level of social interaction in general. The available data doesn't contain much of an explanation. It could be interesting to investigate to what extent sociocultural habits influence the communication patterns and integration of scientists into international scientific communities.

References

Barjak, F., & Harabi, N. (2004). *Internet for research: SIBIS topic report no. 2 to the European Commission, DG information society.* Retrieved January 28, 2005, from http://www.fhso.ch/pdf/unternehmer/sibis_final.pdf

Cohen, J. (1996). Computer mediated communication and publication productivity among faculty. *Internet Research: Electronic Networking Applications and Policy, 6*(2/3), 41-63.

Cole, J. R. & Cole, S. (1973). *Social stratification in science.* Chicago; London: University of Chicago Press.

Crane, D. (1972). *Invisible colleges. Diffusion of knowledge in scientific communities.* Chicago; London: University of Chicago Press.

Crawford, S. Y., Hurd, J. M., & Weller, A. C. (1996). *From print to electronic: The transformation of scientific communication.* Medford, NJ: Information Today Inc.

Cronin, B. (1982). Invisible colleges and information transfer: A review and commentary with particular reference to the social sciences. *Journal of Documentation, 38,* 212-236.

Dubrovsky, V., Kiesler, S., & Sethna, B. (1991). The equalization phenomenon: Status effects in computer-mediated and face-to-face decision making groups. *Human-Computer Interaction, 6,* 119-146.

Finholt, T. A., & Olson, G. M. (1997). From laboratories to collaboratories: A new organizational form for scientific collaboration. *Psychological Science, 8,* 28-36.

Fry, J. (2004). The cultural shaping of ICTs within academic fields: Corpus-based linguistics as a case study. *Literary and Linguistic Computing, 19,* 303-319.

Gläser, J. (2003). What internet use does and does not change in scientific communities. *Science Studies, 16,* 38-51.

Hilgartner, S. (1995). Biomolecular database—New communication regimes for biology? *Science Communication, 17,* 240-263.

Kling, R., & Callahan, E. (2002). *Electronic journals, the internet, and scholarly communication* (CSI Working Paper WP01-04). Retrieved August 5, 2004, from http://www.slis.indiana.edu/csi/WP/wp01-04B.html

Kling, R., & McKim, G. (2000). Not just a matter of time: Field differences and the shaping of electronic media in supporting scientific communication. *Journal of the American Society for Information Science, 51,* 1306-1320.

Lazinger, S.S., Bar-Ilan, J., & Peritz, B.C. (1997). Internet use by faculty members in various disciplines: A comparative case study. *Journal of the American Society for Information Science, 48,* 508-518.

Mitra, A., Hazen, M. D., LaFrance, B., & Rogan, R. G. (1999). Faculty use and non-use of electronic mail: Attitudes, expectations and profiles. *Journal of Computer-Mediated Communication, 4*(3). Retrieved August 5, 2004, from http://www.ascusc.org/jcmc/vol4/issue3/mitra.html

Mulkay, M. (1977). Sociology of the scientific research community. In I. Spiegel-Rösing & D. de Solla Price (Eds.), *Science, technology and society: A cross-disciplinary perspective.* (pp. 93-148). London; Beverly Hills, CA: Sage.

National Telecommunications and Information Administration [NTIA] & Economic and Statistics Administration [ESA] (2000). *Falling through the net: Toward digital inclusion. A report on American's access to technology tools.* Washington: US Department of Commerce. Retrieved on August 3, 2004, from http://search.ntia.doc.gov/pdf/fttn00.pdf

Nentwich, M. (2003). *Cyberscience—Research in the age of the internet.* Vienna: Austrian Academy of Sciences Press.

OECD (2001). *Understanding the digital divide.* Paris: OECD.

Price, D. J. de Solla & Beaver, D. deB. (1965). Collaboration in an invisible college. *American Psychologist, 21,* 1011-1018.

Walsh, J. P., Kucker, S., Maloney, N., & Gabbay, S. (2000). Connecting minds: CMC and scientific work. *Journal of the American Society for Information Science, 51,* 1295-1305.

Walsh, J. P., & Roselle, A. (1999). Computer networks and the virtual college. *STI Review, 24,* 49-77.

Work Research Centre (2003). *Benchmarking social inclusion in the information society in Europe and the US* (Final Report of the SIBIS project). Brussels: European Commission.

Endnotes

[1]   The invisible college has been defined as the "power group of everybody who is really somebody in a field" (Price & Beaver, 1965, p. 1011). An

invisible college controls research resources and decides on the research strategies in its field. It serves as a channel for the dissemination of research ideas and research results which it has evaluated positively. It also represents a regulator that matches the volume of information with the absorptive capacities of the researchers and it is a source of research ideas (Cronin, 1982).

2   Funding from the European Commission within the IST Programme (IST-2000-26276) is gratefully acknowledged.

3   Multivariate analyses showed that some of the gender differences regarding internet use are not primarily caused by gender, but they are rather a consequence of the lower recognition and position of female scientists.

4   Professional recognition was assessed through assessing whether the respondent had won any scientific awards, served on a major professional committee, the editorial board of a scientific journal or a national/international advisory committee.

5   In multivariate analyses it could be corroborated that the differences of e-journal use between research managers, senior and junior researchers are significant, even though the data shown in Table 2 do not really indicate this.

6   Intensity of usage was assessed for the following communication channels: e-mails sent and received, phone calls made and received, letters sent and received, formal face-to-face meetings, informal face-to-face meetings and participation in chat room sessions and video conferences.

Chapter XII

Gender Stratification and E-Science:
Can the Internet Circumvent Patrifocality?

Antony Palackal
Loyola College of Social Sciences, India

Meredith Anderson
Louisiana State University, USA

B. Paige Miller
Louisiana State University, USA

Wesley Shrum
Louisiana State University, USA

Abstract

Can the internet improve the lot of women in the developing world? This study investigates the degree to which the internet affects the constraints on women pursuing scientific careers. We address this question in the context of the scientific community of Kerala, India, developing a "circumvention" argument that fundamentally implicates information and communication technologies in shaping gender roles. We begin by reviewing two main constraints identified in prior research (educational and research localism) that increase the likelihood of restricted professional networks. Next, we examine the extent to which women scientists have gained access to e-science technologies. With evidence of increased access, we argue that the presence of connected computers in the home has increased consciousness of the importance of international contacts. We conclude by proposing that internet connectivity is helping women scientists to circumvent, but not yet undermine, the patrifocal social structure that reduces social capital and impedes career development.

Introduction

E-science technologies have been widely heralded as an equalizer for communities of knowledge workers remote from global scientific centers. They promise to promote development by decreasing the cost and increasing the efficiency of both international and local communication, improving access to information and facilitating international collaboration. Davidson, et al. (2002) refer to such arguments promoting the internet as an "elixir" that will heal developmental woes. What is missing throughout most of this discussion is gender, particularly in reference to the issue of whether the internet differentially affects men and women scientists. Likewise, throughout the vast literature on sex differences in scientific attainment (Cole & Zuckerman, 1984, 1987; Fox, 1995; Fox & Long, 1995; Keller, 1995; Kyvik & Teigen, 1996; McElrath, 1992; Mukhopadhyay, 1994; Ranson, 2003; Wajcman, 1991, 1995; Xie & Shauman, 1998), the impact of new information and communication technologies on the careers of female scientists has largely been neglected.

The main objective of this essay is the examination of qualitative evidence for a specific idea regarding the relationship between gender inequity and e-science technologies under the social structural conditions of patrifocality that characterize most of the Indian subcontinent. We propose that Indian women scientists have begun to use the internet to circumvent gender codes that govern behavior and limit access to social capital, particularly international professional contacts. The consequences of an affirmative answer would be significant for two reasons. First, since enhanced connectivity and access to the internet might then be expected to reduce gender inequities in the knowledge sector. But perhaps more important, what might now be simple technological leverage could begin a process that would undermine the broader social structure of patrifocality.

We develop this argument through an examination of ninety qualitative interviews with agricultural and environmental scientists in Kerala, India. The main purpose is to explore the internet as an equalizing mechanism for women in science in view of major constraints on the development of social capital previously identified as limiting career trajectories for women in Africa and India. We begin by considering gender inequality and the social structural context of patrifocality in the Indian context. We summarize prior work on Kerala women scientists and the characteristics of educational and organizational localism that constrain the development of social capital. Following a description of our methodology, we introduce evidence that women scientists are still highly constrained in their educational and subsequent work experiences. However, qualitative evidence seems to suggest two social changes are occurring: (1) dramatic improvement in access to new information and communication technologies, (2) the importance of international professional ties are becoming part of feminine consciousness. In the last section of the chapter we discuss the implications of these findings in relation to the broader gender structure and maintain that the increasingly common presence of women in the Indian labor force is also functioning to undermine the stringent system of gender stratification.

Science and Gender Inequality

One of the few empirical studies to address the conditions of women scientists in India and Africa is Campion and Shrum's (2004) analysis of 293 scientists in Kenya, Ghana and Kerala (India). Based on data gathered in 1994, when connections to the internet were still extremely rare, this study addressed the question: Why do women in the developing world have more difficulty pursuing research careers than their male counterparts? Through the analysis of gender differences in educational and personal backgrounds, research productivity, professional activities, organizational resources and professional networks, as well as the consequences of these differences, they argued that gender inequality results from a lack of social rather than material resources. On a great variety of indicators, gender differences are not large, but rather trivial or nonexistent. Yet in comparison to their male colleagues, female scientists in Kerala were found to possess disadvantages in terms of participation rates on government committees and editorial boards, publications in international journals, and contacts outside of the local research system, including those with professionals in developed countries. Overall, the networks of male scientists were found to have broader range than those of their female counterparts, leading the authors to suggest that more international opportunities for women should be made available.[1] Our study begins with this background and examines questions of technological and social change, for what has happened in the decade since 1994 is a major shift in the way science is conducted, owing to the widespread diffusion of the internet.

Inequities in the Indian scientific community result from a stringent system of gender roles, combined with the ethnocentric tendency for modeling research careers on those of the developed world (Campion & Shrum, 2004). Subrahmanyan noted that much of the existing literature on the masculine nature of science and the resulting instances of particularism are based on western experiences and social order (1998).[2] One reason career advancement is problematic in the Indian academic sector is a quota system based on caste. Promotion consider-ations are based partly on merit, but often a qualified individual of the appropriate caste must fill a faculty vacancy in order to maintain the prescribed balance within each department. For this reason, university careers often stagnate regardless of individual achievements. Caste issues affect both men and women.

Our focus here is a second overarching feature of the Indian system, which has been described as "patrifocality." While extended discussions may be found in Mukhopadhyay and Seymour (1994) and Subrahmanyan (1998), the significant features are summarized by Gupta and Sharma in their analysis of Indian women scientists:

...subordination of individual interests to the welfare of the family; gender-differentiated family roles with females being associated with the 'private' sphere; gender differentiated family authority structure (with authority of same-generational males over socially equivalent females, such as husbands over wives, brothers over sisters); family control of marriage arrangements; patrilineal descent, inheritance, and succession; patrifocal residence, with daughters shifting allegiance to husband's family after marriage; and an ideology of 'appropriate' female behaviour that emphasizes chastity, obedience, and modesty. (2002, p. 902)

Patrifocality denotes a general set of social structural characteristics that tend to appear together in the Indian social context. To say that a system is patrifocal is not to say that it is without variability, nor is it to say that all of its features are inextricably tied together. For example, in the State of Kerala, the focus of our study, matrilocal residence is not uncommon, but patrifocality still describes the social structural conditions that obtain. For our purposes, the central point is that for Indian women scientists, the patrifocal system involves an extreme commit-ment to localism, restricting interactional opportunities out of micro-level familial concerns for the purity and labor of females. In brief, (1) women should first attend to their caretaking obligations, and (2) their movements out of the household should ordinarily be limited to those required for the fulfillment of those duties. The social limitations implied by patrifocality are clearly pronounced. Gender inequity in educational, travel and work opportunities results in lower access to social capital and embedded network resources (Lin, 2001).

Campion and Shrum's evidence in the specific context of Kerala is part of a broader set of concerns with women's social networks. Gupta and Sharma (2002) maintain that the social networks of Indian women scientists are limited for several reasons. First, their networks are basically gender homophilous due to social segregation in which informal interaction with unrelated men is problematic. Second, their networks are more locally oriented due to cultural constraints upon their geographic mobility. Finally, they lack female mentors and colleagues with whom they can safely collaborate. For these reasons, Indian women are likely to be professionally isolated. As one academic chemical engineer articulates: "I am the only woman in the department. It has a male culture. Men have their own tea clubs. I feel different and isolated. The colleagues talk very little with me" (2002, p. 907).

In this sense, female scientists can be seen to possess a lower degree of social capital than their male counterparts, for their network ties are usually homophilous by sex. Smith-Lovin and McPherson (1993) theorize basic processes by which homophilous networks ensue from a lifetime of socialization. While their work is focused on developed countries, network theory may be employed generally to explain "how small, seemingly inconsequential differences between males and females in childhood or early adulthood can be transformed over the life course into dramatic levels of gender segregation and inequality" (1993, p. 223). Gender homophilous ties to childhood playmates coupled with tendencies to respond differently to network structures result in dissimilar social realities, and therefore different locations in the social structure for girls and boys during their formative years. Women tend to occupy network positions that facilitate the flow of information about the private sphere of household and family, while those of men contain resources that promote upward mobility in the public sphere of career and finance. Ultimately, networks of adult women become densely connected and contain more ties to kin and neighborhood, while networks of adult men are sparser but more extensive, containing more ties to coworkers and outside members. We focus on the consequences of this process in the distinctive context of the Indian system of gender stratification. Our work centers on the distinction between face-to-face contact and electronic communication and information flows made possible by the internet. While we find it premature to regard social change as extensive—indeed, specific evidence continues to be hard to find—we examine the degree to which women scientists have begun to use electronic communication to circumvent patrifocal interactional restrictions, gaining increased access to social networks and social capital.

The Kerala Context

The state of Kerala in southwestern India has long been a focus of attention for the Indian and international development communities. Initially, Kerala garnered renown owing to its unique political and sociocultural history: for instance, the first elected Communist government in the world and matrilineal kinship organization among dominant communities like the Nayars. More recently, interest has been focused on its paradoxical pattern of growth, often referred to as the "Kerala Model of Development," characterized by high social achievements on a weak economic base (George, 1993; Oommen, 1992, 1999; Ramachandran, 1996; Shrum & Iyer, 2000). In the context of the growing incidence of crimes against women, coupled with rising rates of suicide and mental illness, serious doubts have arisen regarding the widely publicized high status of women in Kerala as measured by conventional indicators like literacy and sex ratio. However, within Indian feminist literature, the concept of "status" itself has been criticized as failing to recognize the unequal relations of power between men and women in a society that situates women in an inferior position inside and outside the private sphere (Saradamoni, 1994). Our analysis must be sensitive to these unique conditions. Nonetheless, it is appropriate to ask whether, given the restrictions patrifocality places on women in Kerala as well as other Indian states, new information and communication technologies are affecting gender inequities. Can the internet be regarded as an equalizer?

Methodology

We base the circumvention argument on 90 face-to-face interviews conducted in Kerala between 2003 and 2004 with professional scientists in a variety of fields, emphasizing agriculture, environment and natural resource management. Our informants worked in four scientific organizations (two research institutes and two universities) in or near Thiruvananthapuram, the capital. One of the research institutes operates under the auspices of the central government in New Delhi, while the other is an agency of the state government in Kerala. The size of the two universities required us to select a subset of departments in the agricultural and basic scientific fields. One institution focused almost exclusively on agricultural training, while the other was more comprehensive. We sought interviews with all active scientific staff in the institutes but selectively oversampled women in the universities. This sampling strategy does not allow us to generalize to the entire population of scientists in Kerala. However, it does allow us to explore in detail the ways in which the internet is beginning to affect the scientific community in the context of gender roles.

From 1994 to 2000 the proportion of women scientists in Kerala has remained relatively constant: slightly more than one third, or 37% in 2000 (Sooryamoorthy & Shrum, 2004).[3] Out of 90 respondents from the four organizations, approximately one quarter was women. The national research institute employs a much higher proportion of women than the state institute, where we could speak with only two women. Men comprised a disproportionate number of respondents in all locations. While the percentage of women working in the professional field of science is not equal to that of men, it is much higher than in many developing areas.[4]

Interviews were preceded by the review of each respondent's curriculum vitae and recent publications. Interviews were transcribed and imported into QSR NVivo, a software package designed for the analysis of qualitative data. The primary use of the software was for thematic coding, which allowed the data to be analyzed using Boolean searches, locating passages possessing a feature or combination of features such as intersections or unions (Richards, 1999). The themes or "nodes" were coded according to the scheme summarized in Appendix A. For this analysis, we utilized seven "parent" nodes: Location of Higher Education, Organizational Involvement, Visits Abroad, Professional Contacts, Internet Use and Access, Research-Related E-mail Transactions and Gender Discourse. Each parent node was coded into a number of "child" nodes, which served to organize responses in terms of the diffusion of internet technologies and their impact on female scientists' access to social resources. Searches of the coded interviews were conducted using these nodes, categorized by the sex of the respondent. In what follows we first consider the two sources of localism, education, and travel, previously identified as constraints on social capital. Next, we examine connectivity and access to the internet by men and women scientists, including the role of the computer in the domestic setting. We then turn to the question of professional networking and role circumvention.

Higher Education

The extent and location of higher education are important indicators of gender inequity within the scientific communities of the world. To the extent that women do not have the same *level* of training as men, and to the extent that they do not have access to the same *places*[5] of training as men, gender differences will persist with respect to the human and social capital necessary for the pursuit of scientific research and careers. Sooryamoorthy and Shrum (2004) showed that within the community of agricultural and environmental scientists of Kerala, the percentage of doctorates remained approximately constant from 1994 to 2000. However, our unpublished data shows that women have made significant gains during the same period. While the proportion of males with PhDs went down from 89% to 78%, the proportion of women with doctorates rose from 67% to 76%. While some of this difference—particularly the lower proportion of male PhDs—may be the result of alternative sampling strategies (described in Sooryamoorthy and Shrum, 2004), the increase in female educational qualification is likely to represent real gains.[6]

Location of doctoral degree represents, in a stark sense, opportunity: opportunities to develop social capital, to receive training unavailable in Kerala, to broaden scientific horizons and finally, to acquire prestige. Using 1994 data, Campion and Shrum (2004) found that male scientists in the developing world were more likely to receive training in developed countries, specifically in the United States and Europe, and spent more time abroad for training and education. However, Indian scientists were less likely to receive advanced training abroad than their African counterparts. Sooryamoorthy and Shrum showed that in 2000, only about one in 20 scientists from these institutions received their training in a developed country. Generally, training "outside" Kerala means education in Tamil Nadu or other locations in the Indian subcontinent.

In our qualitative interviews, we examined the persistence of these patterns. All of the academic faculty we interviewed possessed the PhD degree. However, women were slightly more likely to possess doctorates than men within the government research institutes. The reason is not that women are more highly educated in the Kerala context, but that the state research institute, which employs the fewest doctorates, is also the most likely to employ men.

Our analysis of the location of highest degree, and particularly for doctoral studies, indicates that women scientists were more likely to be constrained in terms of location. In comparison to men, they were less likely to receive their degrees from universities outside Kerala. Many female respondents echoed similar sentiments in regard to limited mobility and the barriers placed on them in choosing to pursue higher studies. The vast majority of women completed their higher degrees, and in many cases *all* of their degrees, within the state of Kerala. One Head of Department described her situation as follows:

We cannot move out freely after the household chores. My husband is not the kind who is supportive. . . . My husband is very orthodox. He would get irritated with studying after marriage. He wanted me to have job but did not like me studying after marriage.

Only one woman received a foreign doctorate. In this rare case, she resided in South Korea for three years while her husband and child remained in India.

Women scientists did not generally aspire to foreign degrees—it was simply not part of their thinking. On the contrary, many Indian women reported beginning their doctoral studies as a result of having no definite career plans, or beginning studies only after procuring permission from their husbands. The situation of male scientists is quite different. For men, choice of location of higher education seemed to be a non-issue. Their mobility was uninhibited, allowing them to enroll in the institution of their choice. No male respondent spoke of turning down opportunities to study abroad due to family concerns. On the other hand, one woman recalled interrupting her studies when her husband accepted a fellowship in Canada. When women went abroad, usually their male partners or siblings first undertook foreign assignments and the women, as wives or sisters, accompanied them. Women simply did not make independent decisions to go abroad for studies—educational decisions were inextricably linked to the decisions of other family members.

We conclude that while there are not important differences in the level of training between men and women, the location of higher education (especially the doctorate) was fundamentally constrained by marital and domestic consider-ations. This "educational localism" is consistent with studies by both Gupta and Sharma (2002) and Campion and Shrum (2004). Listening to the stories of women scientists, it seemed clear that women were not in a position to independently choose an academic center for their higher learning, especially after marriage. As most women in southern India are married around the age at which they complete their master's degree, they generally undertake higher degrees with little or no intention of building a career or pursuing one that would detract from their family responsibilities. Gender roles, marital values, and family obligations impinge on "centrifugal" movements (away from family), serving as a set of normative and functional restrictions. Far from a state of equality, the stratified sociocultural milieu of Kerala is such that men have relative freedom of choice, leading to geographic mobility in educational outcomes. For women such freedom is incidental to familial ideology and requirements. Additionally, the element of safety in an alien geographical environment is viewed as an important consideration involved in choosing the location of higher education (Gupta & Sharma, 2002). For this reason, women tend to secure the company of a male spouse or relative before considering study outside of India.

Although these observed trends are not surprising, given the patrifocal social context of India, their implications are especially important to understand in the new context of e-science technologies. Since some internet users receive their computer training in the developed world, women can be seen as possessing an additional disadvantage. Not only are they more likely to receive all of their training from the same institution, limiting their exposure to new approaches and ideas, they are also less likely to benefit from training in the technologically savvy institutions of the west. Furthermore, both male and female respondents who studied or trained outside of India reported that their professional and personal contacts in these countries remained strong and that these individuals continued to serve as partners in collaboration, providers of much needed resources and friends.[7] In this sense, a locally oriented education serves to limit one's access to both social and material resources.

Travel

The second constraint on social capital identified by Campion and Shrum is the gender difference in time spent away from the employing organization (2004). This variety of localism is organizational or "research" related. It is not that women never leave their organizations for workshops, conferences, training and research, but that they spend less time doing it—an average of two weeks less per year. As in educational localism, this results in reduced opportunities for the development of professional linkages that may be important to the scientific career.

Notwithstanding constraints resulting from the patrifocal value system, many female scientists in 2003-2004 have undertaken multiple short-term visits abroad. Most of the senior women scientists and professors who were our informants had traveled abroad for career related activities including short-term training programs, conferences, workshops, seminars and symposia. Women from the government research institutes tended to undertake more foreign visits than did their university counterparts. This observed trend appears to result from the greater tendency of government research institutions to send their employees abroad for training rather than individual motivation. Paper presentations at academic seminars or conferences, particularly those at the international level, were regarded as especially important for career advancement. These international opportunities are more plentiful than those that involved teaching or conducting research abroad. More important, they are brief, which is extremely significant in a context where the domestic labor of women is tightly controlled. Women are much more likely to accept opportunities to visit foreign centers of training when they do not involve extended absences.

In contrast to their female counterparts, most men had undertaken foreign visits, often multiple visits, and these trips tended to last for longer periods of time. Furthermore, more male scientists engaged in foreign travel for the purpose of post-doctoral fellowships, research and employment than their female colleagues. Women scientists, however, tended to conduct visits abroad quite differently, and were found to participate only in the academic program for which the visit was organized, returning to their families immediately afterwards. Male scientists, on the other hand, often took time off for other academic and nonacademic activities when embarking on short-term international travel. In light of the relatively continuous familial obligations of the patrifocal social milieu of Kerala, women are likely to attend conferences, workshops and other short meetings, and less likely to undertake training, collaboration or research projects that would require extended absences from the domestic setting.

These interviews lead us to conclude that the high priority of family obligations affects the frequency, type and duration of visits abroad for Indian women scientists. Campion and Shrum found that male scientists traveled more extensively, arguing that they were therefore more likely to gain exposure to international standards and practices of research (2004). In the earlier time period, nearly half the women in their sample had no experience at all in a foreign country. Our qualitative interviews ten years later yielded little evidence of change: A slight majority of respondents reported at least one visit to the developed world for the purpose of research or training. The frequency may have increased, but what seems stark is the evidence for increased consciousness of the importance of travel.

The importance of travel for Indian scientists cannot be overemphasized, for these visits often result in the creation of network ties with scientists in the developed world. In turn, these ties can later serve as a source of the social and material resources necessary to provide information and enhance productivity. But we note that contacts in the developed world can also be seen to benefit Indian women scientists in another way—in the global scientific community, individuals are less likely to subscribe to the patrifocal ideology that restricts contact between unrelated men and women. The cultivation of these contacts is extremely important to the career of Indian women scientists. But if women are still unable to travel outside the organization apart from infrequent and brief periods, then they are denied the opportunity to develop ties that can effectively subvert patrifocal restrictions on their behavior. As with educational localism, organizational localism based on a duality of self- and socially-imposed restrictions, restricts the face-to-face interactions that have traditionally led to increases in the size and range of scientific networks. The question, then, is whether electronic interactions provide a way of circumventing these restrictions.

Connectivity and Access

E-science technologies are all based on the assumption that researchers are connected to the internet. But as a forthcoming study by Ynalvez, et al., shows, in the context of developing areas (2005), internet connectivity is anything but an nonproblematic concept: While the vast majority of scientists describe themselves as current e-mail users, far fewer have personal computers, ready access to the technology, use it in diverse ways or have extensive experience. In brief, internet "adoption" cannot be characterized as a single act on the part of users. Computer access, bandwidth, technological privacy and the skill required for the effective use of e-science technologies are equally important for scientific research and careers.

One definition of adequate connectivity that has been employed in developmental arguments is a dedicated computer system for each scientist, with relatively high connection speeds, and adequate maintenance, that is, brief periods of downtime (Shrum, 2005). This e-scientific environment is the standard for Western scientists—indeed, it is assumed by most scientists, explicitly or implicitly, when they consider or engage in collaboration across national boundaries.

It was quickly apparent that government research centers were much better provided for in terms of internet connectivity than academic scientists. This finding is consistent with our quantitative assessments of the Kerala research system (Sooryamoorthy & Shrum, 2004; Ynalvez et al., 2005) and it rapidly became a crucial point in our understanding of e-science in Kerala. The disparity between governmental research centers and academic institutions in terms of internet connectivity was pronounced. While both sectors reported the existence of internet connections, only those respondents employed at the two research centers felt they had adequate access to the internet at work. In these locations, there were generally two to three computers in each lab. Respondents often shared computers and many did not claim to have an exclusive connection at work. Both research centers expressed satisfaction with the LAN connections recently acquired by their institutes, but those working at the national institute did complain about the lack of access outside of office hours, which required special permission.

Although the connectivity is for 24 hours [but] we are able to access it only till office hours. At 3.30 in the afternoon...they switch off the main server. Because there are occasions when you need to use it even after office hours--a couple of hours more.

While this may be perceived as a disadvantage, it is more likely to result in gender equity than inequity, since restrictions on the movement of women largely prevent them from working during nonstandard hours—they have fewer opportunities than men to use the facilities outside of the normal working period.

University faculty encountered other issues. Although most have come to terms with the present arrangement of internet facilities, there are simply too few computer systems available to the staff. The number of computers connected to the internet varied by department, rarely exceeding two or three machines. As Ynalvez, et al., report (2005), nearly thirteen individuals shared a single computer in the Kerala academic sector in the year 2000, the highest average in a study that included Ghana and Kenya. While many respondents felt that they would acquire additional connections within the near future, over the five years since we began work there progress has been extremely slow. Not only are there few machines, but their presence in a nonrestricted academic environment has an inevitable consequence: University scientists also had to cope with the fact that students generally monopolized existing connections. According to one marine biologist, "Here in the department only one computer is connected to the internet. So all the students will be crowding around that all the time…teachers don't go and use it during office hours."

Although internet access was available in the libraries and computer labs of both universities, faculty rarely used it, owing to the number of machines and competition with students. The speed and reliability of connections was another common complaint, since many of departments still rely on dial-up connections over undependable telephone lines. This put scientists at a particular disadvantage with respect to downloading the large files sometimes needed for data manipulation or literature review. In sum, a situation exists in which universities have devolved the responsibility for connectivity to the faculty themselves. If they are to use ICTs in a significant way for research and teaching purposes, they must purchase their own computer and provide for a connection. However, it was apparent that the costs are sufficiently low and the demand sufficiently high that nearly all university scientists obtained connections at their homes, where they carried out the majority of their communication and information search activities.

As we consider the domestic use of ICTs to be critical to gender issues, it is important to stress the role of the connected computer within the home and familial setting, where sharers of a single computer are only one quarter as many as they are at work (from 12.7 to 3.2) and parents have authority over its use. The clear majority of scientists in both universities and government research institutes claimed that the internet was a useful tool for professional and personal communication as well as research. Many also spoke of the improvement in the quality of internet connections available in Kerala and most reported regular use and at least basic knowledge of e-mail and literature searching. Where did they acquire this knowledge? Many learned the internet from their children—women were especially likely to report this influence of kids. An agricultural chemist began using the internet in 1999:

We first bought a computer at home. Then we chat regularly with my husband and my sons are there to help me. Emailing I need only when it is required. But chatting we do....My son [created an account]... .Even literature searching he does for me.

The head of a department of plant breeding was asked if anyone was helping her learn the internet:

My daughter...and we have [the connection] in our department since about a year... .It was difficult as I do not know typing. When I do something when I try to open or close a window if something happens I would get stuck. My daughter would help me and then I have a relative close to my house who helped me out.

Several men also mentioned the influence of their sons and daughters on the decision to purchase a home computer and the instruction they provided once the system had been installed. Interestingly, the two respondents who claimed to possess the highest degree of knowledge of internet use, however, were men who gained this knowledge while conducting research in more developed countries.

Professional Networks

The relative disadvantage of women in education and travel should be diminished to the extent that the creation and maintenance of social ties are possible through new ICTs, that is, to the degree that the internet serves as a functional substitute for face-to-face contact. In our qualitative interviews, we asked specifically about the professional contacts that were important to these scientists. The answers to these questions did not yield concrete conclusions, but intriguing possibilities regarding what is or may be happening to the professional networks of respondents within the past decade. The baseline is provided by Campion and Shrum's finding that gender inequities are not pronounced for local professional contacts (2004). Especially important is their evidence that (a) women scientists had larger local networks than men in Kerala; and (b) male scientists had more extensive contacts in India, not including Kerala.

Our recent interviews revealed Indian women were quick to mention various professional contacts throughout the world. In comparison with male scientists, Kerala women mentioned far fewer contacts *within* their departments, and even within India. Instead, they were far more likely to describe their associations with various international scientists. One noted an instance in which she could continue the professional relationship with a scientist from Madras University whom she had met at a conference. In the course of her professional correspondence with him through e-mail, she recalled, with a great sense of pride, approaching him for a fruitful discussion on plant diseases. On another occasion she received a rare sample from an internationally renowned scientist working in a foreign research centre. Following the improvement of connectivity at her institute:

[My] contact has become a lot, really, otherwise contact was very less. Even when I go for training there also I get some help, otherwise we do not communicate... .We have stopped letters. We are ladies, you know, either phone or e-mail, letters are very rare.

This utterance that "we are ladies" is a loaded phrase in the context of patrifocal structure of Kerala society: The clarity and enthusiasm with which she discussed the use of the internet for professional networking is revealing of the ways in which electronic communication is viewed as a social leveller, bringing this younger scientist opportunities she had not experienced.

In contrast, Indian men reported extensively on contacts and collaborations within their departments, institutes and regional areas. Although Indian women employed in the government sector mentioned more local male contacts than their university counterparts, many of these men occupied supervisory positions and were expected to provide feedback on their work. Similarly, the local male contacts of female respondents were more often identified as lab partners, former professors or dissertation advisors than scientists who sought them out on the basis of previous publications or for the purpose of collaboration. Contacts of these types were generally not used for collaboration, but rather for examination purposes or settling questions.

Many female respondents claimed that the internet was responsible for the cultivation of their international ties. Several mentioned submitting papers to international journals or conferences about which they received information online. Also common were accounts of publishing papers in online journals and subsequently receiving e-mail correspondence from interested readers. In a few cases, such exchanges eventually led to visits abroad and international collaboration. Additionally, nearly all participants in the study claimed that e-mail was their preferred mode of communication with all of their contacts, although some expressed the preference of sending letters by way of post to those professional contacts who were also friends. Overall, the utility of e-mail as a tool to enhance productivity was widely accepted despite the fact that it was also widely held to be more impersonal than communication by telephone or post.

The importance of these findings is their implication that Indian women in this study are relying more on international contacts than on contacts at the local level. This finding contrasts with those of Campion and Shrum (2004) as well as those of Gupta and Sharma (2002) with respect to the local orientation of Indian women's social networks. Indeed, if confirmed, this would represent a noteworthy change in the nature and pattern of the social networks of Indian women scientists, especially since the patrifocal sociocultural milieu seems to be firmly intact. Essentially, women seem to be using the internet in order to subvert patrifocal regulations on their mobility and behavior. The internet allows them to create and maintain international network ties to a greater degree than their male colleagues in spite of their lesser opportunities for international travel.

It is generally assumed that the primary reason external contacts are important is for the exposure to new ideas, information and international standards and practices of research. This is true regardless of whether the scientist is from the United States, Africa, or India. But for those embedded in a culture of patrifocality, there are other reasons to value external professional ties. For Indian women, collaboration with male colleagues is often problematic owing to the cultural practices of gender segregation that limit interaction between unrelated men and women to certain kinds of situations. Conversely, professionals in the developing world hold few such beliefs and may be more likely to value the input and experience of Indian women scientists. If we begin to find, as we suspect from these qualitative interviews, that Indian women are relying more heavily on international contacts, it may be because their input is valued more highly by foreign professionals than their own male colleagues. We return to this point in the conclusion.

Women and Work

If internet use has raised consciousness of the importance of international ties, and women are increasingly allowed to make short term visits, can this be the beginning of broader social change? That is, can new ICTs be implicated in the subversion of patrifocality? We cannot address this question directly with the available data, but it seems doubtful that the small developments documented here will go *beyond* circumvention to subversion without further increases in the participation of women in scientific careers. Our interviews yielded extensive reports of women in regard to the historically pervasive limitations on their physical mobility and interactional opportunities. But we want to leave open the possibility that increasing numbers of women pursuing careers in science combined with the ability of the internet to circumvent many of the social and religious codes that govern their behaviour could result in a new reality, at least for women scientists. The internet itself may allow them to acquire the level of social capital necessary to build a successful career without appearing to violate the social reality of gender stratification.

Despite the existence of widely held social perceptions of women as intellectually inferior, our informants, both men and women, were unanimous that no marked difference exists between the sexes with regard to the quality of academic performance. In their own experience with students, participants agreed that males and females perform equally well in studies and research. Some respondents even felt that female students fare better academically and are more committed to academic programs than their male counterparts. One senior scientist at the national research institute aired this sentiment, common to the scientific community, regarding the academic performance of female and male students:

I don't think that there is any conspicuous difference between their programs, particularly in the academic output. Intellectually, both groups are doing well. No difference at that level. When it comes to field activities, because of the practical consideration, boys are in [a] bit more advantageous position—going in ship, cruise, travel etc... .

But this level of academic performance does not serve to impress authorities and supervisors with the need to provide equal interactional opportunities for women scientists that would undermine patrifocal restrictions on physical mobility. Girls remain at a severe disadvantage when it comes to extracurricular activities, such as field visits and library or laboratory work that often extend well into the night. They are seldom allowed to take part in such activities, as their physical mobility remains under the control of their husband or male family members. Many of our respondents readily cited these cultural restrictions when asked about gender differences in performance of scientists. According to Chanana (2001), this seemingly paradoxical reality results from the intersection of macro-level development policies aimed at promoting education among women and girls with micro-level familial concerns of preserving the purity of female family members. A university marine biologist aptly summarized the contradictory nature of the manifestation of such beliefs:

Indian women are very different from their counterparts abroad and the armory of men.... If you ask a female student to be here after five in the evening, do you think that the student will be here?.... You could label as cultural and you are so cautious about yourself and the society and you never let the women work according to their will or from the childhood itself she is groomed in such a way that she has restrictions around her. If you go to a foreign lab you will find a girl working up to 12 in the night and she walks off after that and she comes at 6 in the morning like men.

After marriage, a woman assumes a large amount of household responsibility, which often interferes with her education. One university scientist recalled that her social position as a married woman had led to difficulties and missed opportunities earlier in her career:

Since my husband was also away I had to look after both the children. I liked to attend the summer or winter courses organized by other universities, but I cannot go, as I do not have any help at home.

In this context, Derne (1994) maintains that the limitations placed by marriage on education and training are nearly inevitable. Although men in Kerala generally value higher education for women, they push their female relatives to marry by their early 20s. Such action is viewed as a preventative measure for maintaining one's reputation and respectability. Within the social context of Kerala, this ideology prevails, as women are more highly educated than in any other part of India, but are ultimately expected to marry and lead traditional lives. In fact, within Malayalee culture, arranged marriages are generally preferred to "love marriages" in which the spouse is personally selected. Arranged marriages are considered highly desirable, even for the highly educated. These gender social-ization practices insure that Indian women find career advancement increasingly difficult after marriage. Female respondents in this study confirmed that, as Indian women in general are not in a position to attain the degrees of productivity exhibited by male colleagues, their career trajectories will not be as smooth or far reaching. As we indicated above, owing to increasing educational attainment by women scientists, this is not due to lower levels of education but to its timing and location.

Discussion

Evidence from qualitative interviews with south Indian scientists on their education, travel, connectivity and social networks leads us to propose a "circumvention" hypothesis concerning technological change. New information and communication technologies are used to circumvent the prevailing social structure of patrifocality that restricts interactional opportunities owing to concerns for female purity and control over domestic labor. The current study supports prior work (Sooryamoorthy & Shrum, 2004; Ynalvez et al., 2005) showing that internet connectivity in Kerala has improved dramatically in the past decade. On the positive side, the diffusion of e-science technologies throughout the subcontinent now indicates a relatively high degree of internet access for women scientists. A conservative interpretation of our results would emphasize that small changes that *may* have occurred in terms of professional resources and opportunities available to Indian women have not significantly changed the social and cultural codes that govern female behavior and gender relations. Our analysis does not indicate that the careers of female scientists have become less locally oriented in terms of major and routine interactional opportunities. The one area that may have seen some improvement is the frequency of short-term visits abroad.

Regarding local *orientation*, however, our evidence yielded a different picture and some reason for optimism. The puzzle is clarified by the notion of circumvention. Women were found to possess a virtually unchanged degree of educational localism although they expressed a heightened awareness of international professional contacts and delighted in describing the international travel and professional networking in which they had engaged. Women clearly *discussed* foreign linkages more than their male counterparts and may be engaging in more short-term travel abroad. We cannot establish with this analysis that women have more international professional contacts than men—but it may be equally significant if they *value their international contacts more than men*. As we indicated above, the social structure of interaction between Indian men and women may mean that Indian women value ties with foreign professionals more highly for two reasons: the knowledge exchanged and the style of interaction. In light of these results, it would be too pessimistic to conclude that patrifocal constraints have remained constant. While limits on mobility remain strong, the opportunities to create and maintain professional ties represented by new ICTs represent a circumvention that may herald a relaxation of constraints. Although women are not permitted to travel away from their families for long periods of time, brief, work-related travel may be increasingly acceptable. While the direction of causality is impossible to establish, it seems likely to us that increases in internet-related social capital have actually *encouraged* short-term travel through the establishment of contacts and a greater awareness of conferences and workshops.

Despite the fact that Malayalee culture endorses educational opportunities for women, they are still expected to assume the role of dutiful wife and mother. New information and communication technologies are the primary reason for the decrease in localism. First, many of our respondents report learning about international opportunities on the internet. Second, the probable increase in the range of women's social networks can be attributed to e-mail communication, and an increase in international travel that also bears some relationship to internet communication and information search. Third, there is a perceived decrease in isolation of Indian women scientists, now that they can publish in international journals and online publications, which may increase their visibility throughout the rest of the world. When viewed in the patrifocal structural context, the relationship between female scientists in India and the internet is nontrivial.

In light of the growing importance of non-exploitative development programs, the reduction of gender inequities in research careers is an important aspect of egalitarian science and technology development policies (Harding, 1995). Career attainment and productive research capabilities are limited for women, since many originate from religious prescriptions and limit their physical and social mobility through concerns for female purity and the demand for control over their activities (Abraham, 2000). However, educational attainment for the female scientific community of Kerala has increased over the past decade, such that the proportion of women with doctorates in our sample equals that of men. The problem goes both ways: Merely increasing the number of women will never equalize the scientific playing field, without attention to structurally based factors such as the allocation of resources and access to social capital (Fox, 1995). But technological resources that provide opportunities for social capital development will have limited benefit without other changes that limit career advancement.

In conclusion, we find some reason for believing that the professional careers of Indian women are changing with the diffusion of e-science technologies. In the new context of equal educational levels, but continued limitations of mobility, women scientists have become less locally oriented through the use of new ICTs and may be beginning to develop their international professional networks. This study should be viewed as a starting point for a systematic analysis of professional and organizational networks, to determine whether changes in local orientation are accompanied by real changes in network size. Further, the relationship between age, connectivity and locality deserves further consideration, given the importance of younger cohorts in routinizing internet collaboration (Shrum, 2005). In the early years of the new millennium, there is still reason to hope that circumvention can lead to broader social change.

References

Abraham, M. (2000). *Speaking the unspeakable*. New Brunswick: Rutgers University Press.

Campion, P. A., & Shrum, W. (2004). Gender and science in development: Women scientists in Ghana, Kenya and India. *Science, Technology, and Human Values, 29*(4), 459-485.

Chanana, K. (2001). Hinduism and female sexuality: Social control and education of girls in India. *Sociological Bulletin, 50*(1), 37-63.

Cole, J. R., & Zuckerman, H. (1984). The productivity puzzle: Persistence and change in patterns of publication of men and women scientists. In M. W. Steinkamp & M. L. Maehr (Eds.), *Women in science* (pp. 217-258). Greenwich; London: JAI Press.

Cole, J. R., & Zuckerman, H. (1987). Marriage, motherhood, and research performance in science. *Scientific American, 256*(2), 119-125.

Davidson, T., Sooryamoorthy, R., & Shrum, W. (2002). Kerala connections: Will the internet affect science in developing areas? In B. Wellman & C. Haythornthwaite (Eds.), *The internet in everyday life* (pp. 496-519). Malden, MA: Blackwell.

Derne, S. (1994). Arranging marriages: How fathers' concerns limit women's educational achievements. In C. C. Mukhopadhyay & S. Seymour (Eds.), *Women, education, and family structure in India* (pp. 83-101). San Francisco: Westview Press.

Fox, M. F. (1995). Women in scientific careers. In S. Jasanoff, G. Markle, J. Peterson, & T. Pinch (Eds.), *Handbook of science, technology, and society* (pp. 205-233). Newbury Park: Sage.

Fox, M. F. (1999). Gender, hierarchy and science. In J. S. Chafetz (Ed.), *Handbook of the sociology of gender* (pp. 441-458). New York: Kluwer Academic/Plenum Publishers.

Fox, M. F., & Long, S. J. (1995). Scientific careers: Universalism and particularism. *Annual Review of Sociology, 50*, 45-71.

George, K. K. (1993). *Limits to kerala model of development: An analysis of fiscal crisis and its implications*. Thiruvananthapuram: Centre for Development Studies.

Gupta, N., & Sharma, A. K. (2002). Women academic scientists in India. *Social Studies of Science, 32*(5-6), 901-915.

Harding, S. (1995). Just add women and stir? In Gender working group: UN commission on science and technology for development, *Missing Links* (pp. 295-307). New York: UNIFEM.

Keller, E. F. (1995). The origin, history, and politics of the subject called 'gender and science': A first person account. In S. Jasanoff, G. Markle, J. Peterson, & T. Pinch (Eds.), *Handbook of science, technology, and society* (pp. 80-94). Newbury Park: Sage.

Kyvik, S., & Teigen, M. (1996). Child care, research collaboration, and gender differences in scientific productivity. *Science, Technology, and Human Values, 21*(1), 54-71.

Lin, N. (2001). *Social capital*. Cambridge: University Press.

McElrath, K. (1992). Gender, career disruption, and academic rewards. *The Journal of Higher Education, 63*(3), 269-281.

Mukhopadhyay, C. C. (1994). Family structure and Indian women's participation in science and engineering. In C. C. Mukhopadhyay & S. Seymour (Eds.), *Women, education and family structure in India* (pp. 103-32). Boulder: Westview Press.

Mukhopadhyay, C. C., & Seymour, S. (1994). *Women, education and family structure in India*. Boulder: Westview Press.

Oommen, M.A. (1992). *The Kerala economy*. New Delhi: Oxford & IBH Publishing Company.

Oommen, M. A. (Ed.). (1999). *Kerala's development experience* (Vols. 1 & 2). New Delhi: Concept Publishing.

Ramachandran, V. K. (1996). On kerala's development achievements. In J. Dreze & A. Sen (Eds.), *Indian development: Selected regional perspectives* (pp. 205-356). Delhi: Oxford University Press.

Ranson, G. (2003). Beyond "gender differences": A Canadian study of women's and men's careers in engineering. *Gender, Work, and Organizations, 10*(1), 22-41.

Richards, L. (1999). *Using NVivo in qualitative research*. London: Sage.

Saradamoni, K. (1994). Women, kerala and some development issues. *Economic and Political Weekly, 29*(26 February), 501-509.

Shrum, W. (2005). Reagency of the internet, or how I became a guest for science. *Social Studies of Science, 36*, 1-36.

Shrum, W., & Campion, P. (2000). Are scientists in developing countries isolated? *Science, Technology and Society, 5*(1), 1-34.

Shrum, W., & Iyer, S. R. (2000). Knowledge, democratization, and sustainability: The Kerala 'model' of scientific capacity building. In G. Parayil (Ed.), *Kerala: The development* experience (pp. 157-177). London; New York: Zed Books.

Smith-Lovin, L., & McPherson, M. (1993). You are who you know: A network approach to gender. In P. England (Ed.), *Theory on gender: Feminism on theory* (pp. 223-251). New York: Aldine de Gruyter.

Sooryamoorthy, R., & Shrum, W. (2004). Is Kerala becoming a knowledge society? Evidence from the scientific community. *Sociological Bulletin, 53*(2), 207-221.

Subrahmanyan, L. (1998). *Women scientists in the third world: The Indian experience.* New Delhi: Sage.

Wajcman, J. (1991). *Feminism confronts technology.* University Park: The Pennsylvania State University Press.

Wajcman, J. (1995). Feminist theories of technology. In S. Jasanoff, G. Markle, J. Peterson, & T. Pinch (Eds.), *Handbook of science, technology, and society* (pp. 189-204). Newbury Park: Sage.

Xie, Y., & Shauman, K. A. (1998). Sex differences in research productivity: New evidence about an old puzzle. *American Sociological Review, 63*(6), 847-870.

Ynalvez, M., Duque, R., Sooryamoorthy, R., Mbatia, P., Palackal, A., & Shrum, W. (2005). When do scientists "adopt" the internet? Dimensions of connectivity in developing areas. *Scientometrics, 63*(1), 39-67.

Endnotes

[1]   Gender differences in professional communication networks should be viewed in light of the more general finding that scientific networks in the developing world are always predominantly local (Shrum & Campion, 2000).

[2]   We have not included an account of the feminist critique of science and technology, the use of gendered metaphors in scientific discourse or biased standards of evaluation (see Fox, 1999 for a review).

[3]   The 2000 survey was conducted at the same organizations we examine in this chapter and is used throughout as a point of comparison, with the clear understanding that some aspects have changed in the three intervening years.

[4]   Kerala is the only state in the Indian subcontinent where the ratio of women to men is above one (1,058 females to 1,000 males). Female literacy and education have achieved surprisingly high levels compared to the national average. Total literacy is 90.92%, with male literacy at 94.20% and female literacy at 87.86%.

[5]   In most developing areas, the receipt of "foreign" degrees is considered superior both in terms of quality of education. Our argument concerns the social rather than the human capital implications of "outside" credentials— that is, networks rather than knowledge. In either case, the prestige of an international degree is nearly always higher, which is why we focus specifically on location as an indicator of the process.

6   Since the second survey was conducted in the same institutions where our qualitative interviews were conducted, differences in the level of educational qualifications can be regarded as insignificant for the analysis that follows.

7   It is important to emphasize that this was not independently verified, and other evidence suggests the weakening of such ties over time (Shrum, 2005).

Appendix A

Thematic Organization Used in Interview Coding

- LOCATION OF HIGHER EDUCATION
    - Post Graduation
        - Institute
    - Doctoral Studies
        - institute
        - period of commencement
        - area of study
        - findings of the study
    - Post Doctoral Study
        - institute
        - period of commencement
        - area of study
        - findings of the study
    - Others
- ORGANIZATIONAL INVOLVEMENT
    - Professional
    - Non Professional
    - No Involvement
- VISITS ABROAD
    - Training/Workshops
    - Fellowship
    - Guest Faculty
    - Obtaining Degrees/Taking Courses
    - Conferences
    - Deputation
    - Resource Person
    - Effect/Influence
    - No Professional Visits Abroad
- PROFESSIOAL CONTACTS
    - Purpose
        - professional
        - nonprofessional
    - Origin of the Contact
    - Level of Contact
    - Mode of Communication
- INTERNET USE AND ACCESS
    - Impression About Internet
    - Latest Internet Activity
    - Problems Encountered
    - Unique Internet Experiences
    - Introduction to Computers
    - Impact of Internet
        - teaching
        - social life
        - research
    - Use of Internet
        - research
        - teaching
        - others
    - Browsing
        - duration
        - sites visited
        - ways of browsing
    - Access to Internet
        - institute
            - speed
            - availability
            - future developments in internet access

Appendix A (cont.)

- home
  - speed
  - availability
- cafe
  - speed
- Negative Impact of Internet
- EMAIL
  - Email Activity/Duration Rate
  - Nature of Email Response
  - Correspondence Built on Email
  - Rate of Mails sent and Received
  - Email Transactions
    - partners
      - personal
      - professional
        - research related
        - others
- GENDER ASPECTS
  - Research
  - Occupation
  - Study

About the Authors

Christine Hine is senior lecturer in the Department of Sociology, University of Surrey, UK. Her main research centres on the sociology of science and technology, including ethnographic studies of scientific culture, information technology and the internet. Her published work on research methods and the internet includes *Virtual Ethnography* (Sage, 2000) and the edited collection *Virtual Methods* (Berg, 2005). She is currently exploring the sociology of cyberscience, as recipient of an ESRC research fellowship. A book on the deployment of information and communication technologies in biological systematics is in preparation in connection with that project. Hine has recently been elected president of the European Association for the Study of Science and Technology (http://www.easst.net).

*   *   *

Meredith Anderson is a doctoral student at Louisiana State University in Baton Rouge, USA. Her research interests include the role of information and communication technologies in Third World development policies, particularly those involving women.

Franz Barjak currently works as a research fellow in the Management Department, University of Applied Sciences Solothurn Northwestern, Switzerland. After graduating from the Technical University in Munich with a diploma in geography, he was a member of the research team on regional and urban economics at the Institute for Economic Research Halle (IWH), Germany. His current research focuses on the use of the internet in science and technology transfer, networks in the life sciences and regional firm clusters.

Anne Beaulieu is senior researcher with the Royal Netherlands Academy of Arts and Science, Amsterdam. She received both her BA (humanistic studies) and MA (communication) from McGill University (Canada), and her PhD (science and technology dynamics) from the University of Amsterdam. She is a visiting fellow of the Science Studies Centre, University of Bath, UK, and also has an appointment to the Amsterdam School of Communication Research, University of Amsterdam. She has studied the development and consequences of biomedical digital imaging and databasing technologies, including an ethnographic study of brain imaging. She has written about the intellectual agenda and methodological issues in current ethnographic research on the internet. Currently, she is investigating the development of online infrastructure for knowledge production in women's studies in The Netherlands.

Geoffrey C. Bowker is executive director, Regis and Dianne McKenna professor, Center for Science, Technology and Society, Santa Clara University, USA. He has written, along with Leigh Star, a book on the history and sociology of medical classifications (*Sorting Things Out: Classification and Practice;* MIT Press, September 1999). This book looks at the classification of nursing work, diseases, viruses and race. His latest book, *Memory Practices in the Sciences,* is about formal and informal recordkeeping in science over the past 200 years (MIT Press, 2005). More information, including a number of publications can be found at his web site: http://epl.scu.edu/~gbowker

Bertram C. Bruce is a professor of library and information science, curriculum & instruction, bioengineering, writing studies, and the Center for East Asian & Pacific Studies at the University of Illinois at Urbana-Champaign, USA. Before moving to Illinois, he taught computer science at Rutgers (1971-1974) and was a principal scientist at Bolt Beranek and Newman (1974-90). He received a BA in biology from Rice University (1968) and a PhD in computer sciences from The University of Texas at Austin (1971). His central interest is in learning—the constructive process whereby individuals and organizations develop as they adapt to new circumstances. This work draws on ideas such as John Dewey's theory of inquiry, as well as on action research and situated studies. Much of it has focused on changes in the nature of knowledge, community, and literacy, as discussed in his recent book, *Literacy in the Information Age: Inquiries into Meaning Making with New Technologies,* and other recent writing.

Alexandre Caldas joined the Oxford Internet Institute, University of Oxford, UK, in April 2004 as a faculty research fellow. He is also invited teacher in new media, information and knowledge systems at the London School of Economics, Department of Media and Communications (Media@LSE). He holds a PhD in science and technology policy studies from SPRU, University of Sussex. His dissertation focused on webmetrics, internet indicators, research collaboration and electronic networks. He completed his master's with distinction in economics and management of science and technology at the technical university of Lisbon, Portugal. In addition to a strong quantitative background in academia (first degree in economics), he has considerable practical experience, including executive director of the science and technology park, Abrantes, Portugal; coordinator of the Regional Internet Project, Ribatejo Digital, Portugal (2002-2004), and ICT coordinator for the Science and Technology Foundation, Portuguese Ministry of Science and Technology (1994-2000). Alexandre has taught at Atlantic University and the Technical University in Portugal and has collaborated on a number of research projects for the UK Department of Trade and Industry, the Ministry of Defense, and Portugal's Innovation Agency. His work is focused on e-science, networks and webmetrics and internet indicators. He is a member of the Institute of Electric and Electronic Engineers (IEEE) and the Computer Society.

Catelijne Coopmans is a research associate at the Innovation Studies Centre, Tanaka Business School, Imperial College London and is about to receive her PhD in management studies from the University of Oxford. She has an interdisciplinary background with a strong emphasis on science and technology studies (STS). Her research is on the social dimensions of the development and use of visualization technologies—particularly in the contexts of medicine and engineering design. She has done ethnographic research on medical imaging in a Dutch hospital and with a British image-processing company. She has worked on a DTI funded e-science project and contributed to a scoping study for the ESRC on e-social science (by Steve Woolgar) in 2003.

Beate Elvebakk is a postdoctoral fellow at the Centre for Technology, Innovation and Culture at the University of Oslo, Norway, and researcher at the Institute of Transport Economics, Oslo. She holds an MA in science, technology and society from the University of Oslo and the University of Maastricht, an MA in philosophy and a PhD in technology, innovation, and culture, both from the University of Oslo. Her present project deals with the digitalization of academic journals. She is especially interested in the intersections between technology studies and epistemology and theories of science.

Jenny Fry works as a researcher at the Oxford Internet Institute, University of Oxford, UK, on the Oxford e-Social Science Project studying legal, ethical, institutional and disciplinary barriers to e-science. Jenny has been studying computer-mediated communication and collaboration from a social science perspective for a number of years. Her work has been mainly concerned with disciplinary differences in shaping the appropriation of information and communication technology infrastructures. She has also developed novel web-based methodologies for studying scholarly communities and intellectual fields online. She received her PhD in information science in 2003 from the University of Brighton and has been a postdoctoral research fellow at the Royal Netherlands Academy of Arts and Sciences, Amsterdam and in the School of Information and Library Science at the University of North Carolina, Chapel Hill.

Caroline Haythornthwaite is associate professor at the Graduate School of Library and Information Science, University of Illinois at Urbana-Champaign, USA. Her research examines what kinds of relations support work and learning communities, and how computer media support such interactions. Studies have examined social networks and media use in co-located academic researchers, distance learners, and interdisciplinary research teams; online communication and community; and processes of collaboration, knowledge co-construction, and technology definition in research teams. Her work appears in *The Information Society, Journal of Computer-Mediated Communication, New Media & Society,* and *Information, Communication and Society.* She recently coedited with Michelle M. Kazmer *Learning, Culture and Community in Online Education* (2004, Peter Lang), and with Barry Wellman *The Internet in Everyday Life* (2002, Blackwell).

Tony Hey is corporate vice president for Technical Computing with Microsoft Corporation[1] In this role he coordinates efforts across Microsoft to collaborate with the global scientific community. He is a top researcher in the field of parallel computing, and his experience in applying computing technologies to scientific research helps Microsoft work with researchers worldwide in various fields of science and engineering. Before joining Microsoft, Hey worked as head of the School of Electronics and Computer Science at the University of Southampton, where he helped build the department into one of the preeminent computer science research institutions in England. Since 2001, Hey has served as director of the UK's e-science Initiative, managing the government's efforts to provide scientists and researchers with access to key computing technologies. Hey is a fellow of the UK's Royal Academy of Engineering and has been a member of the European Union's Information Society Technology Advisory Group. He has also served on several national committees in the United Kingdom, including committees of the UK Department of Trade and Industry and the Office of Science and Technology. In addition, Hey has advised countries such as China, France, Ireland, and Switzerland to help them advance their scientific agenda and become more competitive in the global technology economy. Hey received the award of Commander of the Order of the British Empire honor for services to science in the 2005 UK New Year's Honours List. Hey is a graduate of Oxford University, with both an undergraduate degree in physics and a doctorate in theoretical physics.

Karen J. Lunsford is an assistant professor of writing at the University of California, Santa Barbara, USA. She primarily teaches scientific and technical writing, as well as graduate courses such as Literacy in the Information Age and Collaborative Learning/Collaborative Writing. Her publications include articles in *Written Communication*, *Computers and the Humanities*, *Journal of Adolescent and Adult Literacy*, and *JoDI: Journal of Digital Information*. Dr. Lunsford's research employs interdisciplinary approaches to understand the writing practices that people engage in within knowledge ecologies, how argument and argumentation are defined in these ecologies, and what roles technologies play in these practices and definitions.

Martina Merz is a senior scientist both at the Observatory on Science, Policy and Society, University of Lausanne, Switzerland, where she is responsible for the research area "Social Studies of Science," and at the Technology and Society Laboratory, EMPA St. Gallen (Switzerland). She also teaches sociology at the University of Luzern. Her research interests center on social studies of science and on gender and science. She is particularly interested in issues pertaining to conceptual scientific practice, to the reconfigurations involved with computer-based research, and to linking up approaches from constructivist science and technology studies with other areas of social science scholarship.

B. Paige Miller is a doctoral student at Louisiana State University in Baton Rouge, USA. Her research focuses on the role information and communication technologies, particularly the internet, play in the research careers of female scientists and academics in developing areas.

Michael Nentwich has been the senior scientist at the Institute of Technology Assessment (ITA) of the Austrian Academy of Sciences in Vienna since 1996. He previously worked at the Vienna University of Economics, the Universities of Warwick and Essex, and the Max Planck Institute for the Study of Societies Cologne. He studied law, political science and economics in Vienna and Bruges and received the *venia docendi* in science and technology studies (STS) in 2004. His present field of research is technology assessment in the field of information and communication technologies. He has published *inter alia,* seven books and over 40 articles, and also serves as the editor of an e-journal and as webmaster.

Antony Palackal teaches in the Post Graduate Department of Sociology in Loyola College of Social Sciences, Kerala, India. He is the India coordinator of the World Science Project that studies the impact of the internet on research communication in developing countries. He obtained his PhD from University of Kerala, India for a doctoral dissertation in the area of globalization and culture. Besides having published many research and popular articles, he authored *Culture, Resistance and Spirituality* in the regional language, and coauthored *Managing Water and Water Users—Experience from Kerala.* He has been a consultant and trainer to several NGOs and socio-political movements in Kerala. He offers classes, courses, and seminars on topics of sociocultural significance. Areas of specialization include: sociology of development, social movements, cultural dynamics, and gender and society.

Wesley Shrum is professor of sociology at Louisiana State University, USA, and secretary of the Society for Social Studies of Science (shrum@lsu.edu). Since 1992 he has been engaged in a study of scientific communication in developing areas, focusing on Kenya, Ghana, and the State of Kerala (India). With the spread of the internet through these areas, he shifted focus in the late 1990s to the globalization of science. (Manuscripts may be found at http://worldsci.net/global) For the past five years he has been involved in providing connectivity to various institutes in these locations, in expanding the quantitative and qualitative dimensions of the survey, and in documenting science in developing areas through audiovisual methodologies.

Katie Vann was schooled in faculties of philosophy and social science in the United States; she writes about the politics of resource organization and management and the social logics of method in the human sciences. She is currently employed at the Virtual Knowledge Studio of the Royal Netherlands Academy of Arts and Science.

Steve Woolgar is a sociologist who holds the Chair of Marketing at the University of Oxford. He was formerly professor of sociology, head of the Department of Human Sciences, and director of CRICT at Brunel University. He took his BA (first class honours), MA and PhD from Emmanuel College, Cambridge University. From 1997-2002, he was director of the ESRC Programme

*Virtual Society?—The social science of electronic technologie*s, comprising 22 research projects at 25 universities throughout the UK. He has published widely in social studies of science and technology, social problems and social theory and has been translated into Dutch, French, Greek, Italian, Japanese, Portuguese, Spanish, and Turkish. His latest book is *Virtual society?—Technology, cyberbole, reality* (Oxford University Press, 2002). His current projects include research into the governance and accountability relations of mundane technical solutions to social problems; whether or not STS (science and technology studies) really means business; and an inquiry into the nature and dynamics of provocation.

Paul Wouters is programme leader of the *Virtual Knowledge Studio for the Humanities and Social Sciences*, at the Netherlands Royal Academy of Arts and Sciences, Amsterdam. From 2001 to 2005, he led the research group Nerdi (Networked Research and Digital Information), also at the Academy. Wouters has an MSc in biochemistry and a PhD in science studies. He has written on the history of the Science Citation Index and citation analysis, research evaluation, and science and technology policy. His present interests are revolving around applications and implications of information and communication technologies in research, especially in the humanities and social sciences.

Endnote

[1]   Microsoft is a registered trademark of Microsoft Corp. in the United States and/or other countries. The names of actual companies and products mentioned herein may be the trademarks of their respective owners.

Index

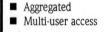
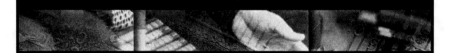

Building a Virtual Library

Ardis Hanson and Bruce Lubotsky Levin
University of South Florida, USA

The organization, functioning, and the role of libraries in university communities continue to change dramatically. While academic research libraries continue to acquire information, organize it, make it available, and preserve it, the critical issues for their management teams in the twenty-first century are to formulate a clear mission and role for their library, particularly as libraries transition to meet the new information needs of their university constituents. *Building a Virtual Library* addresses these issues by providing insight into the current changes and developments within the area of library science.

ISBN 1-59140-106-2(h/c); eISBN 1-59140-114-3 • US$79.95 • 255 pages • Copyright © 2003

"It is critical for the university to make longstanding financial commitments to support the library's role in the academic online environment. This includes innovative funding initiatives and commitments for resources that the library and university together must identify and establish."
–*Ardis Hanson and Bruce Lubotsky Levin*
University of South Florida, USA

It's Easy to Order! Order online at www.idea-group.com or call 1-717-533-8845 x10!
Mon-Fri 8:30 am-5:00 pm (est) or fax 24 hours a day 717/533-8661

 **Information Science Publishing**
Hershey • London • Melbourne • Singapore • Beijing

An excellent addition to your library